VIDEO NOTEBOOK

JUDITH A. BEECHER
JUDITH A. PENNA

ALGEBRA & TRIGONOMETRY
GRAPHS AND MODELS
SIXTH EDITION

PRECALCULUS
GRAPHS AND MODELS,
A RIGHT TRIANGLE APPROACH
SIXTH EDITION

Marvin L. Bittinger
Indiana University Purdue University Indianapolis

Judith A. Beecher

David J. Ellenbogen
Community College of Vermont

Judith A. Penna

PEARSON

Boston   Columbus   Indianapolis   New York   San Francisco
Amsterdam   Cape Town   Dubai   London   Madrid   Milan   Munich   Paris   Montreal   Toronto
Delhi   Mexico City   São Paulo   Sydney   Hong Kong   Seoul   Singapore   Taipei   Tokyo

Reproduced by Pearson from electronic files supplied by the author.

Copyright © 2017 Pearson Education, Inc.
Publishing as Pearson, 501 Boylston Street, Boston, MA 02116.

ISBN-13: 978-0-13-426722-7
ISBN-10: 0-13-426722-2

1 2 3 4 5 6 OPM 20 19 18 17 16

www.pearsonhighered.com

Contents

Section 1.1   Introduction to Graphing

Plotting Points, (x, y), in a Plane

Each point (x, y) in the plane is described by an **ordered pair**. The first number, x, indicates the point's horizontal location with respect to the y-axis, and the second number, y, indicates the point's vertical location with respect to the x-axis. We call x the **first coordinate**, the **x-coordinate**, or the **abscissa**. We call y the **second coordinate**, the **y-coordinate**, or the **ordinate**.

Example 1   Graph and label the points $(-3, 5)$, $(4, 3)$, $(3, 4)$, $(-4, -2)$, $(3, -4)$, $(0, 4)$, $(-3, 0)$, and $(0, 0)$.

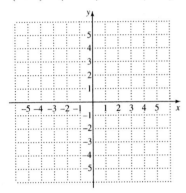

Example 2   Determine whether each ordered pair is a solution of the equation $2x + 3y = 18$.

a) $(-5, 7)$

$$2x + 3y = 18$$

$$2\left(\boxed{}\right) + 3\left(\boxed{}\right) \overset{?}{\mid} 18$$

$$-10 + 21$$

$$\boxed{} \mid 18 \qquad \underline{}$$
$$\text{True / False}$$

$(-5, 7)$ _____ a solution.
    is / is not

b) $(3, 4)$

$$2x + 3y = 18$$

$$2\left(\boxed{}\right) + 3\left(\boxed{}\right) \overset{?}{\mid} 18$$

$$6 + 12$$

$$\boxed{} \mid 18 \qquad \underline{}$$
$$\text{True / False}$$

$(3, 4)$ _____ a solution.
    is / is not

---

x- and y-Intercepts

An **x-intercept** is a point $(a, 0)$. To find a, let $y = 0$ and solve for x.
A **y-intercept** is a point $(0, b)$. To find b, let $x = 0$ and solve for y.

Example 3   Graph: $2x + 3y = 18$.

Find the y-intercept: $\left(0, \boxed{}\right)$

$2 \cdot \boxed{} + 3y = 18$

$0 + 3y = 18$

$3y = 18$

$y = \boxed{}$

Find the x-intercept: $\left(\boxed{}, 0\right)$

$2x + 3 \cdot \boxed{} = 18$

$2x + 0 = 18$

$2x = 18$

$x = \boxed{}$

Find a third point: $\left(5, \boxed{}\right)$

$2 \cdot \boxed{} + 3y = 18$

$10 + 3y = 18$

$3y = \boxed{}$

$y = \boxed{}$

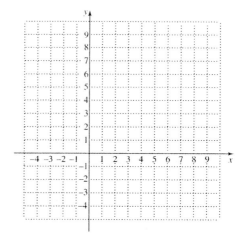

Example 4   Graph: $3x - 5y = -10$.

$3x - 5y = -10$

$-5y = \boxed{} - 10$

$y = \dfrac{3}{5}x + \boxed{}$

When $x = -5$, $y = \dfrac{3}{5}\left(\boxed{}\right) + 2 = -3 + 2 = \boxed{}$.

When $x = 0$, $y = \dfrac{3}{5} \cdot \boxed{} + 2 = 0 + 2 = \boxed{}$.

When $x = 5$, $\dfrac{3}{5} \cdot \boxed{} + 2 = 3 + 2 = \boxed{}$.

x	y	(x, y)
-5	-1	$(-5, -1)$
0	$\boxed{}$	$\left(0, \boxed{}\right)$
5	5	$(5, 5)$

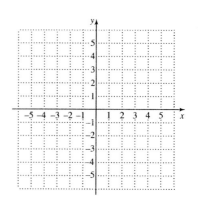

Example 5   Graph: $y = x^2 - 9x - 12$.

Let $x = -3$:

$$y = \left(\boxed{}\right)^2 - 9\left(\boxed{}\right) - 12$$

$$= 9 + \boxed{} - 12 = 24$$

$$\left(-3, \boxed{}\right)$$

Let $x = 2$:

$$y = \boxed{}^2 - 9 \cdot \boxed{} - 12$$

$$= \boxed{} - 18 - 12 = -26$$

$$\left(2, \boxed{}\right)$$

x	y	(x, y)
-3	24	$(-3, 24)$
-1	-2	$(-1, -2)$
0	-12	$(0, -12)$
2	-26	$(2, -26)$
4	-32	$(4, -32)$
5	-32	$(5, -32)$
10	-2	$(10, -2)$
12	24	$(12, 24)$

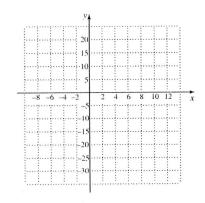

The Distance Formula

The **distance** d between any two points (x_1, y_1) and (x_2, y_2) is given by

$$d = \sqrt{(x_2 - x_1)^2 + (y_2 - y_1)^2}.$$

Example 6   Find the distance between each pair of points.

a) $(-2, 2)$ and $(3, -6)$

$$d = \sqrt{(x_2 - x_1)^2 + (y_2 - y_1)^2}$$

$$= \sqrt{\left[3 - \left(\boxed{}\right)\right]^2 + \left(\boxed{} - 2\right)^2}$$

$$= \sqrt{\boxed{}^2 + (-8)^2}$$

$$= \sqrt{25 + 64}$$

$$= \sqrt{\boxed{}} \approx \boxed{}$$

b) $(-1, -5)$ and $(-1, 2)$

$$d = \sqrt{(x_2 - x_1)^2 + (y_2 - y_1)^2}$$

$$= \sqrt{\left[-1 - \left(\boxed{}\right)\right]^2 + \left(-5 - \boxed{}\right)^2}$$

$$= \sqrt{0^2 + \left(\boxed{}\right)^2}$$

$$= \sqrt{\boxed{}} = \boxed{}$$

Example 7   The point $(-2, 5)$ is on a circle that has $(3, -1)$ as it center. Find the length of the radius of the circle.

$$d = \sqrt{(x_2 - x_1)^2 + (y_2 - y_1)^2}$$

$$d = \sqrt{\left(-2 - \boxed{}\right)^2 + \left(5 - \left(\boxed{}\right)\right)^2}$$

$$= \sqrt{(-5)^2 + \boxed{}^2}$$

$$= \sqrt{\boxed{} + 36}$$

$$= \sqrt{61} \approx 7.8$$

The Midpoint Formula

If the endpoints of a segment are (x_1, y_1) and (x_2, y_2), then the coordinates of the **midpoint** of the segment are

$$\left(\frac{x_1 + x_2}{2}, \frac{y_1 + y_2}{2}\right).$$

Example 8   Find the midpoint of the segment whose endpoints are $(-4, -2)$ and $(2, 5)$.

$$\left(\frac{x_1 + x_2}{2}, \frac{y_1 + y_2}{2}\right)$$

$$= \left(\frac{-4 + \boxed{}}{2}, \frac{\boxed{} + 5}{2}\right)$$

$$= \left(\frac{-2}{2}, \frac{3}{2}\right)$$

$$= \left(\boxed{}, \frac{3}{2}\right)$$

Example 9   The diameter of a circle connects the points $(2, -3)$ and $(6, 4)$ on the circle. Find the coordinates of the center of the circle.

$$\left(\frac{x_1 + x_2}{2}, \frac{y_1 + y_2}{2}\right)$$

$$= \left(\frac{2 + \boxed{}}{2}, \frac{\boxed{} + 4}{2}\right)$$

$$= \left(\boxed{}, \frac{1}{2}\right)$$

The Equation of a Circle

The standard form of the equation of a circle with center (h, k) and radius r is

$(x-h)^2 + (y-k)^2 = r^2$.

Example 10   Find an equation of the circle having radius 5 and center $(3, -7)$.

$(x-h)^2 + (y-k)^2 = r^2$

$h = 3 \quad k = \boxed{} \quad r = 5$

$(x-3)^2 + \left[y - \left(\boxed{}\right)\right]^2 = 5^2$

$(x-3)^2 + \left(y + \boxed{}\right)^2 = \boxed{}$

Example 11   Graph the circle $(x+5)^2 + (y-2)^2 = 16$.

$(x-h)^2 + (y-k)^2 = r^2$

(h, k) center, r radius

$\left[x - \left(\boxed{}\right)\right]^2 + (y-2)^2 = \boxed{}^2$

Center: $\left(-5, \boxed{}\right)$

Radius: $\boxed{}$

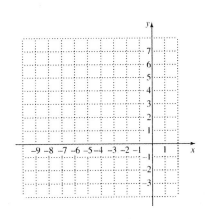

When we graph a circle using a graphing calculator, we select a viewing window in which the distance between the units is visually the same on both axes. This is called **squaring the viewing window**. On many calculators we can do this by choosing a window in which the length of the y-axis is $\dfrac{2}{3}$ the length of the x-axis with Xscl = Yscl. The window with dimensions $[-6, 6, -4, 4]$, $[-9, 9, -6, 6]$, and $[-12, 12, -8, 8]$ are examples of squared windows.

Example 12   Graph the circle $(x-2)^2+(y+1)^2=16$.

We will graph this circle on a graphing calculator using the Circle feature from the DRAW menu. To do this we need to know the center and the radius of the circle.

$$(x-h)^2+(y-k)^2=r^2$$

$$\left(x-\boxed{}\right)^2+\left[y-(-1)\right]^2=\boxed{}^2$$

Center (h,k): $\left(2,\boxed{}\right)$

Radius r: 4

Press $\boxed{2^{\text{nd}}}$ $\boxed{\text{PRGM}}$ to see the DRAW menu. Then use the $\boxed{\nabla}$ key to move to item 9, Circle. Press $\boxed{\text{ENTER}}$ or $\boxed{9}$ to copy "Circle(" to the home screen. Now enter the coordinates of the center followed by the radius by pressing $\boxed{2}$ $\boxed{,}$ $\boxed{}$ $\boxed{,}$ $\boxed{4}$ $\boxed{)}$. Choose a squared window. One good choice is $[-9,9,-6,6]$. Press $\boxed{\text{ENTER}}$ to see the graph.

Section 1.2   Functions and Graphs

> **Function**
>
> A **function** is a correspondence between a first set, called the **domain**, and a second set, called the **range**, such that each member of the domain corresponds to *exactly one* member of the range.

Example 1   Determine whether each of the following correspondences is a function. (Draw the arrows shown in the video as your first step in each case.)

a) −6

6   36

−3   The correspondence _____ a function.

9   is / is not

3

0   0

b)

Appointing President	Supreme Court Justice	
George H. W. Bush	Samuel A. Alito, Jr.	
	Stephen G. Breyer	
William Jefferson Clinton	Ruth Bader Ginsburg	The correspondence _____ a function.
	Elena Kagan	is / is not
George W. Bush	John G. Roberts, Jr.	
	Sonia M. Sotomayor	
Barack H. Obama	Clarence Thomas	

Example 2   Determine whether each of the following correspondences is a function.

a)

Domain	Correspondence	Range
Years in which a presidential election occurs	The person elected	A set of presidents

This correspondence _____ a function because in each presidential election
                              is / is not
exactly one president is elected.

b)

Domain	Correspondence	Range
All automobiles produced in 2016	Each automobile's VIN (Vehicle Identification Number)	A set of VINs

This correspondence _____ a function because each automobile has *exactly one*
VIN.   is / is not

c)

Domain	Correspondence	Range
The set of all professional golfers who won a PGA tournament in 2014	The tournament won	The set of all PGA tournaments in 2014

This correspondence _____ a function because a winning golfer could be
                    is / is not
paired with more than one tournament.

d)

Domain	Correspondence	Range
The set of all PGA tournaments in 2014	The winner of the tournament	The set of all golfers who won a PGA tournament in 2014

This correspondence _____ a function because each tournament has only one
winning golfer.   is / is not

Relation

When a correspondence between two sets is not a function, it may still be an example of a
relation. A **relation** is a correspondence between a first set, called the **domain**, and a
second set, called the **range**, such that each member of the domain corresponds to *at least
one* member of the range.

Example 3   Determine whether each of the following relations is a function. Identify the
domain and the range.

a) $\{(9,-5),\ (9,5),\ (2,4)\}$

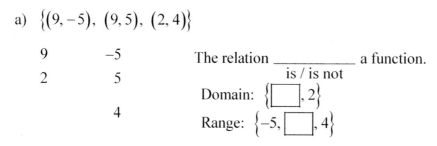

9   −5   The relation _____ a function.
2   5   is / is not
                        Domain: $\left\{\boxed{},2\right\}$
            4   Range: $\left\{-5,\boxed{},4\right\}$

b) $\{(-2,5),\ (5,7),\ (0,1),\ (4,-2)\}$

$\begin{array}{cc} -2 & 5 \\ 5 & 7 \\ 0 & 1 \\ 4 & -2 \end{array}$

The relation _____ a function.
is / is not

Domain: $\left\{-2, 5, 0, \boxed{}\right\}$

Range: $\left\{5, \boxed{}, 1, -2\right\}$

c) $\{(-5,3),\ (0,3),\ (6,3)\}$

$\begin{array}{cc} -5 & \\ 0 & 3 \\ 6 & \end{array}$

The relation _____ a function.
is / is not

Domain: $\left\{-5, \boxed{}, 6\right\}$

Range: $\left\{\boxed{}\right\}$

Example 4   A function f is given by $f(x) = 2x^2 - x + 3$. Find each of the following.

a) Find $f(0)$.

$$f(0) = 2 \cdot \boxed{}^2 - \boxed{} + 3$$
$$= 2 \cdot \boxed{} - 0 + 3$$
$$= 0 - 0 + 3$$
$$= 3$$
$$f(0) = \boxed{}$$

b) Find $f(-7)$.

$$f(-7) = 2 \cdot \left(\boxed{}\right)^2 - \left(\boxed{}\right) + 3$$
$$= 2 \cdot \boxed{} + \boxed{} + 3$$
$$= 98 + 7 + 3$$
$$= 108$$
$$f(-7) = \boxed{}$$

c) Find $f(5a)$.

$$f(5a) = 2\left(\boxed{}\right)^2 - \boxed{} + 3$$
$$= 2\left(\boxed{}\right) - 5a + 3$$
$$= \boxed{} - 5a + 3$$
$$f(5a) = 50a^2 - 5a + 3$$

d) Find $f(a-4)$.

$$f(a-4) = 2\left(\boxed{}\right)^2 - \left(\boxed{}\right) + 3$$
$$= 2\left(a^2 - \boxed{} + 16\right) - (a-4) + 3$$
$$= 2a^2 - 16a + \boxed{} - \boxed{} + 4 + 3$$
$$= 2a^2 - \boxed{} + 39$$
$$f(a-4) = 2a^2 - 17a + 39$$

Example 5   Graph each of the following functions.

a) $f(x) = x^2 - 5$

x	$f(x)$	$(x, f(x))$
-3	4	$(-3, 4)$
-2	-1	$\left(-2, \boxed{}\right)$
-1	$\boxed{}$	$\left(-1, \boxed{}\right)$
0	-5	$\left(\boxed{}, -5\right)$
1	$\boxed{}$	$(1, -4)$
2	-1	$(2, -1)$
3	$\boxed{}$	$\left(3, \boxed{}\right)$

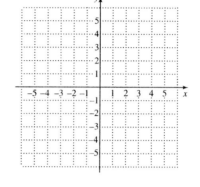

b) $f(x) = x^3 - x$

x	$f(x)$	$(x, f(x))$
-2	-6	$(-2, -6)$
-1	$\boxed{}$	$\left(-1, \boxed{}\right)$
0	0	$(0, 0)$
1	$\boxed{}$	$\left(\boxed{}, 0\right)$
2	6	$\left(2, \boxed{}\right)$
$-\dfrac{1}{2}$	$\dfrac{3}{8}$	$\left(-\dfrac{1}{2}, \dfrac{3}{8}\right)$
$\dfrac{1}{2}$	$-\dfrac{3}{8}$	$\left(\dfrac{1}{2}, -\dfrac{3}{8}\right)$

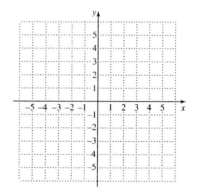

c) $f(x) = \sqrt{x + 4}$

x	$f(x)$	$(x, f(x))$
-4	$\boxed{}$	$(-4, 0)$
-3	1	$\left(-3, \boxed{}\right)$
0	$\boxed{}$	$\left(\boxed{}, 2\right)$
5	3	$(5, 3)$

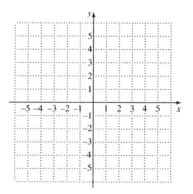

Example 6   For the function $f(x) = x^2 - 6$, use the graph at left below to find each of the following function values.

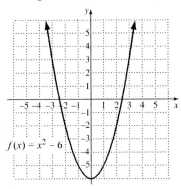

$f(x) = x^2 - 6$

a) Find $f(-3)$.

$$f(-3) = \boxed{}$$

b) Find $f(1)$.

$$f(1) = \boxed{}$$

The Vertical-Line Test

If it is possible for a vertical line to cross a graph more than once, then the graph *is not* the graph of a function.

Example 7   Which of graphs (a) – (f) are graphs of functions? In graph (f), the solid dot shows that $(-1, 1)$ belongs to the graph. The open circle shows that $(-1, -2)$ does *not* belong to the graph.

a)

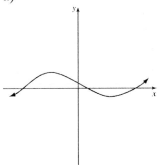

<u>   </u>

Yes / No

b)

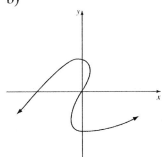

<u>   </u>

Yes / No

c)

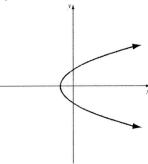

<u>   </u>

Yes / No

d)

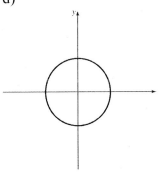

<u>   </u>

Yes / No

e)

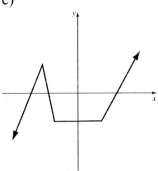

<u>   </u>

Yes / No

f)

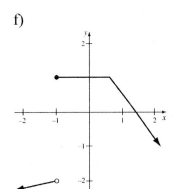

<u>   </u>

Yes / No

Example 8   Find the indicated function values, if possible, and determine whether the given values are in the domain of the function.

a) $f(1)$ and $f(3)$, for $f(x) = \dfrac{1}{x-3}$

$$f(1) = \frac{1}{\boxed{}-3} = \frac{1}{\boxed{}} = -\frac{1}{2}$$

$f(1)$ is defined; 1 $\underline{\hspace{2cm}}$ in the domain of f.
 is / is not

$$f(3) = \frac{1}{\boxed{}-3} = \frac{1}{0}$$

$f(3)$ is not defined; 3 $\underline{\hspace{2cm}}$ in the domain of f.
 is / is not

b) $g(16)$ and $g(-7)$, for $g(x) = \sqrt{x} + 5$

$$g(16) = \sqrt{\boxed{}} + 5 = 4 + 5 = 9$$

$g(16)$ is defined; 16 $\underline{\hspace{2cm}}$ in the domain of g.
 is / is not

$$g(-7) = \sqrt{\boxed{}} + 5; \quad \sqrt{-7} \text{ is not defined as a real number.}$$

$g(-7)$ is not defined; -7 $\underline{\hspace{2cm}}$ in the domain of g.
 is / is not

Example 9   Find the domain of each of the following functions.

a) $f(x) = \dfrac{1}{x-7}$

$$x - 7 = 0$$

$$x = \boxed{}$$

Thus, $\boxed{}$ is not in the domain of f.

Domain: $\left\{x \middle| x \neq \boxed{}\right\}$, or $\left(-\infty, \boxed{}\right) \cup \left(\boxed{}, \infty\right)$

b) $h(x) = \dfrac{3x^2 - x + 7}{x^2 + 2x - 3}$

$x^2 + 2x - 3 = 0$

$(x + 3)\left(x - \boxed{}\right) = 0$

$x + 3 = 0$ or $x - 1 = 0$

$x = \boxed{}$ or $x = \boxed{}$

Thus, -3 and $\boxed{}$ are not in the domain of h.

Domain: $\left\{ x \middle| x \neq -3 \text{ and } x \neq \boxed{} \right\}$, or $\left(-\infty, \boxed{}\right) \cup \left(-3, \boxed{}\right) \cup (1, \infty)$

c) $f(x) = x^3 + |x|$

We can substitute any real number for x.

Domain: $\{x | x \text{ is a real number}\}$, or $\left(-\infty, \boxed{}\right)$.

d) $g(x) = \sqrt[3]{x - 1}$

Because the index is _____, the radicand, $x - 1$, can be any real number. Thus x
odd / even

can be any real number.

Domain: $\{x | x \text{ is a real number}\}$, or $\left(\boxed{}, \infty\right)$.

Example 10   Using the graph of the function, find the domain and the range of the function.

a)

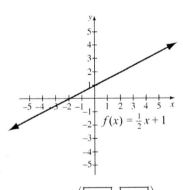

Domain $= \left(\boxed{}, \boxed{}\right)$

Range $= \left(\boxed{}, \boxed{}\right)$

b)

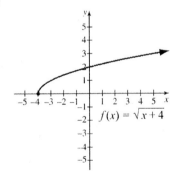

Domain $= \left[\boxed{}, \boxed{}\right)$

Range $= \left[\boxed{}, \boxed{}\right)$

c)

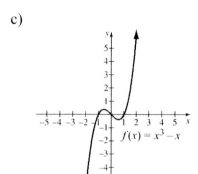

Domain = (☐ , ☐)

Range = (☐ , ☐)

d)

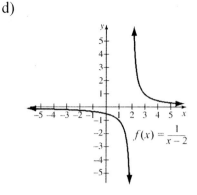

Domain = (☐ , ☐) ∪ (☐ , ☐)

Range = (☐ , ☐) ∪ (☐ , ☐)

e)

$f(x) = x^4 - 2x^2 - 3$

Domain = (☐ , ☐)

Range = [☐ , ☐)

f)

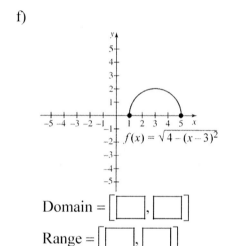

Domain = [☐ , ☐]

Range = [☐ , ☐]

Example 11   The linear expansion L of the steel center span of a suspension bridge that is 1420 m long is a function of the change in temperature t, in degrees Celsius, from winter to summer and is given by

$$L(t) = 0.000013 \cdot 1420 \cdot t,$$

where 0.000013 is the coefficient of linear expansion for steel and L is in meters. Find the linear expansion of the steel center span when the change in temperature from winter to summer is 30°, 42°, 50°, and 56° Celsius.

$$L(30) = 0.000013 \cdot 1420 \cdot \boxed{} = 0.5538 \text{ m,}$$

$$L(42) = 0.000013 \cdot 1420 \cdot \boxed{} = \boxed{} \text{ m,}$$

$$L(50) = 0.000013 \cdot 1420 \cdot \boxed{} = \boxed{} \text{ m, and}$$

$$L(56) = 0.000013 \cdot 1420 \cdot \boxed{} = \boxed{} \text{ m.}$$

Section 1.3   Linear Functions, Slope, and Applications

Linear Functions

A function f is a **linear function** if it can be written as

$\quad f(x) = mx + b,$

where m and b are constants. If $m = 0$, the function is a **constant function** $f(x) = b$. If $m = 1$ and $b = 0$, the function is the **identity function** $f(x) = x$.

Slope

The **slope** m of a line containing points (x_1, y_1) and (x_2, y_2) is given by

$$m = \frac{\text{rise}}{\text{run}} = \frac{\text{the change in } y}{\text{the change in } x} = \frac{y_2 - y_1}{x_2 - x_1} = \frac{y_1 - y_2}{x_1 - x_2}.$$

Example 1   Graph the function $f(x) = -\dfrac{2}{3}x + 1$ and determine its slope.

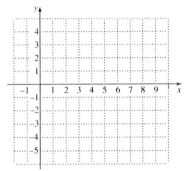

$f(3) = -\dfrac{2}{3} \cdot \boxed{} + 1$

$\quad = -2 + 1 = \boxed{}$

$f(9) = -\dfrac{2}{3} \cdot \boxed{} + 1$

$\quad = -6 + 1 = \boxed{}$

We have pairs $(3, -1)$ and $(9, -5)$.

$$m = \frac{y_2 - y_1}{x_2 - x_1} = \frac{-5 - \left(\boxed{}\right)}{\boxed{} - 3} = \frac{-4}{6} = -\frac{2}{3}$$

It doesn't matter which two points we use to find the slope. We find a third point on the line.

$f(6) = -\dfrac{2}{3} \cdot \boxed{} + 1$

$\quad = -4 + 1 = \boxed{}$

Consider the pairs $(3, -1)$ and $(6, -3)$ and find the slope.

$$m = \frac{\boxed{} - (-1)}{6 - \boxed{}} = \frac{-2}{3} = -\frac{2}{3}$$

This illustrates that it doesn't matter which two points we use to find the slope as long as we do the subtractions in the right order.

Horizontal Lines and Vertical Lines

Horizontal lines are given by equations of the type $y = b$ or $f(x) = b$. (They are functions.) If a line is horizontal, the change in y for any two points is 0 and the change in x is nonzero. Thus a horizontal line has slope 0.

Vertical lines are given by equations of the type $x = a$. (They are *not* functions.) If a line is vertical, the change in x is 0. Thus the slope is *not defined* because we cannot divide by 0.

Example 2   Graph each linear equation and determine its slope.

a) $x = -2$

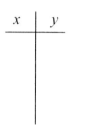

Slope is _____.

$$m = \frac{0-3}{-2-(-2)} = \frac{-3}{\boxed{}}$$

b) $y = \dfrac{5}{2}$

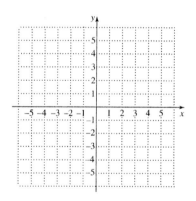

$m = \boxed{}$

Example 3   *Curb Ramps.* Curb ramps provide independent access to sidewalks for those who use wheelchairs. Guidelines for the grade of a curb ramp suggest a grade between 5.9% and 8.3%. A federal law states that every vertical rise of 1 ft requires a horizontal run of at least 12 ft. (*Source:* Federal Highway Administration, Office of Planning, Environment, and Realty) Find the grade of the curb ramp shown in the figure in the video.

$$\text{grade} = m = \frac{3 \text{ in.}}{\boxed{}} = \frac{\boxed{}}{14} \approx 7.1\%$$

Average Rate of Change

Slope can also be considered as an **average rate of change**. To find the average rate of change between any two data points on a graph, we determine the slope of the line that passes through the two points.

Example 4   The number of people participating in the federal Supplemental Nutrition Assistance Program has increased from 17.2 million in 2000 to 47.6 million in 2013. Find the average rate of change in the number of people using food stamps from 2000 to 2013.

We use the points $(2000, 17.2)$ and $(\boxed{}, 47.6)$.

Slope = Average rate of change

$$= \frac{\text{change in } y}{\text{change in } \boxed{}}$$

$$= \frac{47.6 - \boxed{}}{2013 - \boxed{}} = \frac{30.4}{13} \approx 2.3$$

The average rate of change over the 13-year period was an increase of $\boxed{}$ million participants per year.

Example 5   Increased oil production in the United States has resulted in decreased imports of crude oil. The total number of barrels imported in 2008 was 3,590,000. This number decreased to 2,810,000 barrels in 2013. Find the average rate of change in crude oil imports from 2008 to 2013.

We use the points $(2008, 3,590,000)$ and $(2013, \boxed{})$.

Slope = Average rate of change

$$= \frac{\text{change in } y}{\text{change in } x}$$

$$= \frac{2,810,000 - \boxed{}}{\boxed{} - 2008} = \frac{-780,000}{5} = -156,000$$

The average rate of change over the 5-year period was a decrease of $\boxed{}$ barrels per year.

> **The Slope-Intercept Equation**
>
> The linear function f given by
>
> $\quad f(x) = mx + b$
>
> is written in slope-intercept form. The graph of an equation in this form is a straight line parallel to $f(x) = mx$. The constant m is called the slope, and the y-intercept is $(0, b)$.

Example 6   Find the slope and the y-intercept of the line with equation $y = -0.25x - 3.8$.

$$y \quad = \quad mx \quad + \quad b \qquad \text{Slope-intercept equation}$$
$$\qquad\qquad \downarrow \qquad\qquad \downarrow$$
$$\qquad \text{Slope} \qquad (0, b)$$
$$\qquad\qquad\qquad y\text{-intercept}$$

$y = -0.25x - 3.8$

Slope is $\boxed{}$.

y-intercept is $\left(0, \boxed{}\right)$.

Example 7   Find the slope and the y-intercept of the line with equation $3x - 6y - 7 = 0$.

$3x - 6y - 7 = 0$

$\quad -6y - 7 = \boxed{}$

$\qquad -6y = -3x + \boxed{}$

$\qquad \dfrac{-6y}{-6} = \dfrac{-3x + 7}{-6}$

$\qquad\qquad y = \boxed{} \, x - \dfrac{7}{6}$

$\qquad\qquad\qquad \underset{m}{\downarrow} \quad \underset{b}{\downarrow}$

Slope is $\boxed{}$.

y-intercept is $\left(0, \boxed{}\right)$.

Example 8   Graph: $y = -\dfrac{2}{3}x + 4.$

$y = mx + b$

$y = -\dfrac{2}{3}x + 4$

Slope is $\boxed{}$.

y-intercept is $\left(0, \boxed{}\right)$.

$-\dfrac{2}{3} = \dfrac{-2}{3} = \dfrac{2}{-3}$

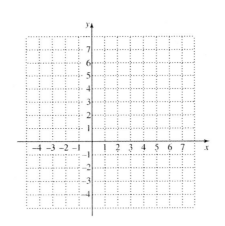

Example 9   There is no *proven* way to predict a child's adult height, but a linear function can be used to *estimate* it, given the sum of the child's parents' heights. The adult height M, in inches, of a male child whose parents' total height is x, in inches, can be estimated with the function

$$M(x) = 0.5x + 2.5.$$

The adult height F, in inches, of a female child whose parents' total height is x, in inches, can be estimated with the function

$$F(x) = 0.5x - 2.5.$$

Estimate the height of a female child whose parents' total height is 135 in. What is the domain of this function?

$F(x) = 0.5x - 2.5$

$F(135) = 0.5\left(\boxed{}\right) - 2.5$

$ = \boxed{} - 2.5 = 65$

$F(135) = \boxed{}$ in., or 5 ft 5 in.

Domain: $(100, 170)$

Section 1.4   Equations of Lines and Modeling

Example 1   A line has slope $-\dfrac{7}{9}$ and y-intercept $(0, 16)$. Find an equation of the line.

$y = mx + b$ Given:

$y = \boxed{}\, x + \boxed{}$ $m = -\dfrac{7}{9}$

$(0, b) = (0, 16)$

Example 2   A line has slope $-\dfrac{2}{3}$ and contains the point $(-3, 6)$. Find an equation of the line.

$y = mx + b$ Given:

$6 = \boxed{}\cdot\left(\boxed{}\right) + b$ $m = -\dfrac{2}{3}$

$6 = \boxed{} + b$ $(x, y) = (-3, 6)$

$4 = b$

$y = -\dfrac{2}{3}x + 4$

Point-Slope Equation

The **point-slope equation** of the line with slope m passing through (x_1, y_1) is

$y - y_1 = m(x - x_1)$.

Example 3   Find an equation of the line containing the points $(2, 3)$ and $(1, -4)$.

Point-slope equation:

$y - y_1 = m(x - x_1)$

$y - \boxed{} = \boxed{}(x - 2)$ $m = \dfrac{-4 - 3}{1 - 2}$

$y - 3 = 7x - \boxed{}$ $= \dfrac{-7}{-1} = 7$

$y = 7x - 11$, or $(x_1, y_1) = (2, 3)$

$f(x) = 7x - 11$

Slope-intercept equation:

$y = mx + b$

$y = \boxed{} x + b$

$-4 = 7 \cdot \boxed{} + b$

$-4 = 7 + b$

$-11 = b$

$y = 7x - \boxed{}$

$m = 7$

$(x, y) = (1, -4)$

Parallel Lines

Vertical lines are **parallel**. Nonvertical lines are **parallel** if and only if they have the same slope and different y-intercepts.

Perpendicular Lines

Two lines with slopes m_1 and m_2 are **perpendicular** if and only if the product of their slopes is -1:

$m_1 m_2 = -1.$

Lines are also **perpendicular** if one is vertical $(x = a)$ and the other is horizontal $(y = b)$.

Example 4   Determine whether each of the following pairs of lines is parallel, perpendicular, or neither.

a) $y + 2 = 5x,$ $5y + x = -15$

$\qquad y = 5x - 2$ $\qquad 5y = -x - 15$

$\qquad m_1 = \boxed{}$ $\qquad y = -\dfrac{1}{5}x - 3$

$\qquad\qquad\qquad\qquad m_2 = \boxed{}$

$m_1 \cdot m_2 = 5\left(-\dfrac{1}{5}\right) = -1$

The lines are _____.
$\qquad\qquad$ parallel / perpendicular / neither

b) $2y + 4x = 8,$ $5 + 2x = -y$

$\qquad 2y = -4x + 8$ $\qquad -y = 2x + 5$

$\qquad\quad y = -2x + 4$ $\qquad\quad y = -2x - 5$

$\qquad\quad m_1 = \boxed{}$ $\qquad\quad m_2 = \boxed{}$

The lines are _____.

$\qquad\qquad$ parallel / perpendicular / neither

c) $2x + 1 = y$ $y + 3x = 4$

$\qquad\quad y = 2x + 1$ $\qquad\quad y = -3x + 4$

$\qquad\quad m_1 = \boxed{}$ $\qquad\quad m_2 = \boxed{}$

The lines are neither parallel nor perpendicular.

Example 5   Write equation of the lines (a) parallel to and (b) perpendicular to the graph of the line $4y - x = 20$ and containing the point $(2, -3)$.

$4y - x = 20$

$\qquad 4y = x + 20$

$\qquad\quad y = \dfrac{1}{4}x + 5$

$\qquad\quad m = \boxed{}$

a) Line parallel to given line:

$\qquad m = \dfrac{1}{4}, \ (2, -3)$

$\qquad\qquad y - y_1 = m(x - x_1)$

$\quad y - \left(\boxed{}\right) = \boxed{}\,(x - 2)$

$\qquad\qquad y + 3 = \dfrac{1}{4}x - \dfrac{1}{2}$

$\qquad\qquad\quad y = \dfrac{1}{4}x - \boxed{}$

b) Line perpendicular to given line:

$$m = -\frac{4}{1}, \text{ or } \boxed{}; \ (2, -3)$$

$$y - y_1 = m(x - x_1)$$

$$y - (-3) = \boxed{}\left(x - \boxed{}\right)$$

$$y + 3 = -4x + 8$$

$$y = -4x + \boxed{}$$

Example 6   The cost of a 30-sec Super Bowl commercial has increased $1.6 million from 2010 to 2015. Model the data in the table with a linear function. Then estimate the cost of a 30-sec commercial in 2018.

Year, x	Cost of a 30-sec Super Bowl Commercial (in millions)
2010, 0	$2.9
2011, 1	3.1
2012, 2	3.5
2013, 3	3.8
2014, 4	4.0
2015, 5	4.5

We use the data points $(1, 3.1)$ and $(4, 4.0)$.

$$m = \frac{4.0 - \boxed{}}{\boxed{} - 1} = \frac{0.9}{3} = \boxed{}$$

Now find the equation of the line.

$$y - y_1 = m(x - x_1)$$

$$y - 3.1 = 0.3\left(x - \boxed{}\right)$$

$$y - 3.1 = 0.3x - 0.3$$

$$y = 0.3x + \boxed{}$$

Now we estimate the cost of a 30-sec Super Bowl commercial in 2018. In 2018,

$$x = 2018 - \boxed{} = 8.$$

$$y = 0.3x + 2.8$$

$$y = 0.3(8) + 2.8 \qquad \text{Substituting}$$

$$y = 2.4 + \boxed{}$$

$$y = 5.2$$

We estimate that the cost of a 30-sec Super Bowl commercial will be $\boxed{}$ million in 2018.

Example 7   Fit a regression line to the data on the cost of a 30-sec Super Bowl commercial given in the table. Then use the function to predict the cost of a 30-sec commercial in 2018.

Year, x	Cost of a 30-sec Super Bowl Commercial (in millions)
2010, 0	$2.9
2011, 1	3.1
2012, 2	3.5
2013, 3	3.8
2014, 4	4.0
2015, 5	4.5

On a graphing calculator, press $\boxed{\text{STAT}}$ and select $\boxed{\text{EDIT}}$ Enter the x-values in L1 and the y-values in L2. To create a scatterplot of the data, first press $\boxed{2^{\text{nd}}}$ $\boxed{y=}$ $\boxed{1}$ (or $\boxed{\text{ENTER}}$). We will use the window $[-1, 6, 0, 8]$. Press $\boxed{\text{GRAPH}}$ to see the scatterplot.

To find the regression line press $\boxed{\text{STAT}}$ $\boxed{\triangledown}$, highlight item 4, and press $\boxed{\text{ENTER}}$. We have

$$y = 0.3142857143x + 2.847619048.$$

Copy the equation to the $y =$ screen as Y1 by pressing $\boxed{\text{VARS}}$ and choosing statistics, then EQ, and finally RegEQ. Now press $\boxed{\text{GRAPH}}$ to see the regression line graphed with the scatterplot. To predict the cost in 2018, evaluate Y1 for $x = 2018 - 2010$, or 8. Press $\boxed{\text{VARS}}$, choose YVARS and then Y1. Finally press $\boxed{(}$ $\boxed{}$ $\boxed{)}$ $\boxed{\text{ENTER}}$ to find Y1(8). We get [   ], so we predict that the cost of a 30-sec commercial will be about $5.36 million in 2018.

Section 1.5   Linear Equations, Functions, Zeros, and Applications

Equation-Solving Principles

For any real numbers a, b, and c:

The Addition Principle: If $a = b$ is true, then $a + c = b + c$ is true.

The Multiplication Principle: If $a = b$ is true, then $ac = bc$ is true.

Example 1   Solve: $\dfrac{3}{4}x - 1 = \dfrac{7}{5}$.

$$\dfrac{3}{4}x - 1 = \dfrac{7}{5} \qquad \text{LCD=20}$$

$$20\left(\dfrac{3}{4}x - 1\right) = \boxed{} \cdot \dfrac{7}{5}$$

$$20 \cdot \dfrac{3}{4}x - 20 \cdot 1 = 20 \cdot \dfrac{7}{5}$$

$$\boxed{}x - 20 = 28$$

$$15x = \boxed{}$$

$$\dfrac{15x}{15} = \dfrac{48}{15}$$

$$x = \dfrac{48}{15}$$

$$x = \boxed{}$$

Check:

$$\dfrac{3}{4}x - 1 = \dfrac{7}{5}$$

$$\dfrac{3}{4} \cdot \boxed{} - 1 \;?\; \dfrac{7}{5}$$

$$\dfrac{12}{5} - \dfrac{5}{5}$$

$$\dfrac{7}{5} \;\Big|\; \dfrac{7}{5} \qquad \text{TRUE}$$

The solution is $\boxed{}$.

Example 2   Solve: $2(5 - 3x) = 8 - 3(x + 2)$.

$$2(5 - 3x) = 8 - 3(x + 2)$$

$$10 - 6x = 8 - 3x - 6$$

$$10 - 6x = 2 - 3x$$

$$10 - 6x + \boxed{} = 2 - 3x + 6x$$

$$10 = 2 + 3x$$

$$10 - 2 = 2 + 3x - \boxed{}$$

$$\boxed{} = 3x$$

$$\dfrac{8}{3} = \dfrac{3x}{3}$$

$$\boxed{} = x$$

Check:

$$2(5 - 3x) = 8 - 3(x + 2)$$

$$2\left(5 - 3 \cdot \boxed{}\right) \;?\; 8 - 3\left(\boxed{} + 2\right)$$

$$2(5 - 8) \;\Big|\; 8 - 3\left(\dfrac{14}{3}\right)$$

$$2(-3) \;\Big|\; 8 - 14$$

$$-6 \;\Big|\; -6 \qquad \text{TRUE}$$

The solution is $\boxed{}$.

Example 3   Solve: $-24x + 7 = 17 - 24x$.

$$24x - 24x + 7 = \boxed{} + 17 - 24x$$

$$7 = 17 \qquad\qquad \text{False equation}$$

This equation has _____ solution.

Example 4   Solve: $3 - \dfrac{1}{3}x = -\dfrac{1}{3}x + 3$.

$$\frac{1}{3}x + 3 - \frac{1}{3}x = \boxed{} - \frac{1}{3}x + 3$$

$$3 = 3 \qquad\qquad \text{True equation}$$

This equation has an _____ number of solutions.
The solution set is $(-\infty, \infty)$.

Example 5   Companies in China purchased 36,560 industrial robots in 2013. This is 54% more than the number of industrial robots purchased by companies in the United States. How many industrial robots were bought by companies in the United States in 2013?

1.   **Familiarize.** Suppose that U.S. companies purchased 20,000 industrial robots. Then the number of robots purchased by Chinese companies would be

$$20,000 + \boxed{}(20,000) = 1.54(20,000)$$

$$= 30,800.$$

Since Chinese companies bought 36,560 robots, we know that the estimate of 20,000 is _____ .
    $\underline{\text{too high / too low}}$

Let $x =$ the number of industrial robots purchased by U.S. companies. Then

$$x + 0.54x = \boxed{} x$$ represents the number of robots purchased by Chinese companies.

2.   **Translate.** We have

$$1.54x = 36,560.$$

3.   **Carry out.** We solve the equation.

$$1.54x = 36,560$$

$$x = \frac{36,560}{1.54} \approx 23,740$$

4. **Check.** 54% of $\boxed{} \approx 12{,}820$ and $23{,}740 + \boxed{} = 36{,}560$. The answer checks.

5. **State.** Companies in the United States purchased approximately 23,740 industrial robots in 2013.

Example 6   In the 2012-2013 school year 47,058 U.S. students studied abroad in Italy and in France. There were 12,638 more students studying in Italy than in France. Find the number of U.S. students studying abroad in Italy and in France.

1. **Familiarize.** Let x = the number of U.S. students studying in France. Then
    $x + \boxed{}$ = the number of U.S. students studying in Italy.

2. **Translate.** We have
    $$x + x + 12{,}638 = \boxed{}.$$

3. **Carry out.**
    $$x + x + 12{,}638 = 47{,}058$$
    $$2x + 12{,}638 = 47{,}058$$
    $$2x = 34{,}420$$
    $$x = \boxed{}$$

    If $x = 17{,}210$, then $x + 12{,}638 = 17{,}210 + 12{,}638 = \boxed{}$.

4. **Check.** $17{,}210 + 29{,}848 = 47{,}058$; also, 29,848 is 12,638 more than 17,210. The answer checks.

5. **State.** In the 2012-2013 school year, there were 29,848 U.S. students studying abroad in Italy and $\boxed{}$ students studying in France.

The Motion Formula

The distance d traveled by an object moving at rate r in time t is given by

$d = r \cdot t.$

Example 7   Delta Airlines' fleet includes B737/800's, each with a cruising speed of 517 mph, and Saab 340B's, each with a cruising speed of 290 mph (*Source*: Delta Airlines). Suppose that a Saab 340B takes off and travels at its cruising speed. One hour later, a B737/800 takes off and follows the same route, traveling at its cruising speed. How long will it take the B737/800 to overtake the Saab 340B?

1. **Familiarize.** Let $t =$ time in hours for B737/800 to overtake the Saab 340B. Then
    $\boxed{} =$ time in hours for Saab 340B to be overtaken.

	d	t	r	
B737/800	d	t	517	$d = 517t$
Saab 340B	d	$t+1$	290	$d = 290(t+1)$

2. **Translate.** The distances are the same, so we have

    $517t = 290(t+1).$

3. **Carry out.**

    $517t = 290t + \boxed{}$

    $227t = 290$

    $t = \dfrac{290}{227} \approx \boxed{}$

4. **Check.**

    B737/800 travels about 1.28 hr; Saab 340B travels $1.28 + 1 = \boxed{}$ hr.

    For B737/800, $d \approx 517(1.28) \approx 661.76$ mi.

    For Saab 340B, $d \approx 290(2.28) \approx 661.2$ mi.

    Since 661.76 mi ≈ 661.2 mi, the answer checks.

5. **State.** About $\boxed{}$ hr after the B737/800 has taken off, it will overtake the Saab 340B.

The Simple-Interest Formula

The **simple interest** I on a principal of P dollars at interest rate r for t years is given by

$$I = Prt.$$

Example 8 Damarion's two student loans total $28,000. One loan is at 5% simple interest and the other is at 3% simple interest. After 1 year, Damarion owes $1040 in interest. What is the amount of each loan?

	P	$\cdot$	r	$\cdot$	t	$=$	I
5% loan	x		☐		1		$0.05x$
3% loan	☐		0.03		1		$0.03(28{,}000 - x)$
Total	28,000						☐

$$I_{5\%} + \quad I_{3\%} = 1040$$

$$\boxed{} + 0.03(28{,}000 - x) = 1040$$

$$0.05x + \boxed{} - 0.03x = 1040$$

$$\boxed{} + 840 = 1040$$

$$0.02x = 200$$

$$x = \frac{200}{0.02} \cdot \frac{100}{100} = \frac{20{,}000}{2} = \boxed{}$$

$10,000 at 5%, $18,000 at 3%

Check: $\$500 + \$540 = \$1040$

Damarian borrowed ☐ at 5% interest and ☐ at 3% interest.

Example 9   In December 2009, a solar energy farm was completed at the Denver International Airport. More than 9200 rectangular solar panels were installed. A solar panel, or photovoltaic panel, converts sunlight into electricity. The length of a panel is 13.6 in. less than twice the width, and the perimeter is 207.4 in. Find the length and the width of a panel.

1. **Familiarize.** Let $w = $ width of a panel. Then $2w - 13.6 = $ length of a panel.

    Perimeter $= \boxed{}$ in.

2. **Translate.**

    $$P = 2l + 2w$$

    $$207.4 = 2\left(\boxed{}\right) + 2\left(w\right)$$

3. **Carry out.**

    $$207.4 = 2\left(2w - 13.6\right) + 2w$$

    $$207.4 = 4w - \boxed{} + 2w$$

    $$207.4 = \boxed{} - 27.2$$

    $$234.6 = 6w$$

    $$39.1 = w$$

    $$2w - 13.6 = 2\left(\boxed{}\right) - 13.6 = 78.2 - 13.6 = \boxed{}$$

4. **Check.** The length 64.6 in. is 13.6 in. less than twice the width 39.1 in.

    Also, $2 \cdot \boxed{} + 2 \cdot 39.1$ in. $= 129.2$ in. $+ 78.2$ in. $= 207.4$ in.

5. **State.** The length of the solar panel is $\boxed{}$ in., and the width is $\boxed{}$ in.

Example 10   Metro Taxi charges a $2.50 pickup fee and $2 per mile traveled. Grayson's cab fare from the airport to his hotel is $32.50. How many miles did he travel in the cab?

1. **Familiarize.** Suppose Grayson traveled 12 mi. Then his fare would be

$$\$2.50 + \$2 \cdot 12 = \$2.50 + \$24 = \boxed{}.$$

The guess is $\underline{}$.
 too high / too low

Let m = the number of miles he travels.

2. **Translate.**

Pickup fee	plus	cost per mile	times	miles traveled	is	total charge
2.50	+	$\boxed{}$	$\cdot$	m	=	32.50

3. **Carry out.**

$$2.50 + 2 \cdot m = 32.50$$
$$2m = \boxed{}$$
$$m = 15$$

4. **Check.** If Grayson travels 15 mi, the mileage charge is $\$2 \cdot 15 = \boxed{}$. The pickup fee is $2.50 and $\$30 + \$2.50 = \boxed{}$. The answer checks.

5. **State.** Grayson traveled $\boxed{}$ mi in the cab.

Zeros of Functions

An input c of a function f is called a **zero** of the function if the output for the function is 0 when the input is c. That is, c is a zero of f if $f(c) = 0$.

Example 11   Find the zero of $f(x) = 5x - 9$.

$$5x - 9 = 0$$
$$5x = 9$$
$$x = \boxed{}, \text{ or } 1.8$$

Section 1.6 Solving Linear Inequalities

Principles for Solving Inequalities

For any real numbers a, b, and c:

The Addition Principle for Inequalities:

If $a < b$ is true, then $a + c < b + c$ is true.

The Multiplication Principle for Inequalities:

a) If $a < b$ and $c > 0$ are true, then $ac < bc$ is true.

b) If $a < b$ and $c < 0$ are true, then $ac > bc$ is true.

   (When both sides of an inequality are multiplied by a negative number, the inequality sign must be reversed.)

Similar statements hold for $a \le b$.

Example 1   Solve each of the following. Then graph the solution set.

a) $3x - 5 < 6 - 2x$ b) $13 - 7x \ge 10x - 4$

$$3x - 5 < 6 - 2x$$

$$3x + \boxed{} - 5 < 6 - 2x + \boxed{}$$

$$5x - 5 < 6$$

$$5x < \boxed{}$$

$$x < \frac{11}{5}$$

$$\left\{ x \,\middle|\, x < \boxed{} \right\}, \quad \left(-\infty, \frac{11}{5} \right)$$

$$13 - 7x \ge 10x - 4$$

$$13 - 7x + \boxed{} \ge 10x - 4 + \boxed{}$$

$$13 \ge 17x - 4$$

$$17 \ge 17x$$

$$\frac{17}{17} \ge \frac{17x}{17}$$

$$\boxed{} \ge x$$

$$\left\{ x \,\middle|\, x \boxed{} 1 \right\}, \quad \left(-\infty, \boxed{} \right]$$

Example 2 Find the domain of the function.

a) $f(x) = \sqrt{x-6}$

$x - 6 \geq \boxed{}$

$x \geq \boxed{}$

Set notation: $\{x \mid x \geq 6\}$

Interval notation: $\left[\boxed{}, \infty\right)$

b) $h(x) = \dfrac{x}{\sqrt{3-x}}$

$3 - x > 0$

$-x > \boxed{}$

$x \boxed{} 3$

Set notation: $\{x \mid x < 3\}$

Interval notation: $\left(\boxed{}, 3\right)$

Example 3 Solve $-3 < 2x + 5 \leq 7$. Then graph the solution set.

$-3 < 2x + 5 \ and \ 2x + 5 \leq 7$

$-3 < 2x + 5 \leq 7$

$-3 - 5 < 2x + 5 - \boxed{} \leq 7 - 5$

$\boxed{} < 2x \leq 2$

$\dfrac{-8}{2} < \dfrac{2x}{2} \leq \dfrac{2}{2}$

$-4 < \boxed{} \leq 1$

The solution set is $\left\{x \mid \boxed{} < x \leq 1\right\}$, $\left(-4, \boxed{}\right]$.

Example 4 Solve $2x - 5 \leq -7 \ or \ 2x - 5 > 1$. Then graph the solution set.

$2x - 5 \leq -7 \quad or \quad 2x - 5 > 1$

$2x \leq -2 \quad or \quad 2x > 6$

$x \leq \boxed{} \quad or \quad x > \boxed{}$

The solution set is $\{x \mid x \leq -1 \ or \ x > 3\}$, $(-\infty, -1] \cup (3, \infty)$.

Example 5   For her house-painting job, Erica can be paid in one of two ways:

   Plan A: $250 plus $10 per hour;

   Plan B: $20 per hour.

Suppose that a job takes n hours. For what values of n is plan B better for Erica?

1. **Familiarize.** Suppose $n = 20$.

   A: $\$250 + \$10 \cdot \boxed{} = \$250 + \$200 = \$450$

   B: $\$20 \cdot \boxed{} = \$400$

   For $n = 20$, Plan A is better.

   Suppose $n = 30$.

   A: $\$250 + \$10 \cdot \boxed{} = \$250 + \$300 = \$550$

   B: $\$20 \cdot \boxed{} = \$600$

   For $n = 30$, Plan B is better.

2. **Translate.**

Income from Plan B	is greather than	income from Plan A
$\boxed{}$	$>$	$250 + 10n$

3. **Carry out.**

   $20n > 250 + 10n$

   $\boxed{} > 250$

   $n > \boxed{}$

4. **Check.**

   A: $\$250 + \$10 \cdot \boxed{} = \$250 + \$250 = \boxed{}$

   B: $\$20 \cdot 25 = \boxed{}$

   When $n = 25$, the income from the plans is the same. In the Familiarize step we saw that Plan B is better for $n = 30$, and $30 > 25$, so our answer seems reasonable.

5. **State.** For values of $n > \boxed{}$ hr, plan B is better for Erica.

Section 2.1   Increasing, Decreasing, and Piecewise Functions; Applications

Example 1   Determine the intervals on which the function is

a) increasing;

b) decreasing;

c) constant.

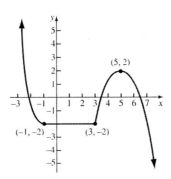

a) As x-values increase from 3 to ☐, y-values increase from ☐ to 2. Thus the function is increasing on $\left(\boxed{}, 5\right)$.

b) As x-values increase from $-\infty$ to -1, y-values _____. Also, as
    increase / decrease

   x-values increase from 5 to ∞, y-values _____. Thus the function is
    increase / decrease

   decreasing on $\left(-\infty, \boxed{}\right)$ and $\left(\boxed{}, \infty\right)$.

c) As x-values increase from -1 to ☐, the y-values stay the same, -2. The function is constant on $\left(\boxed{}, 3\right)$.

Example 2   Use a graphing calculator to determine any relative maxima or minima of the function $f(x) = 0.1x^3 - 0.6x^2 - 0.1x + 2$ and the intervals on which the function is increasing or decreasing.

Enter the function as Y1 and graph it in a window that shows the curvature well. We use $[-4, 6, -3, 3]$.

Use the MAXIMUM and MINIMUM features from the CALC menu to find the relative maximum and the relative minimum.

Relative maximum is about 2.004 when $x \approx$ [   ].

Relative minimum is about [   ] when $x \approx 4.082$.

Increasing: $(-\infty, -0.082), ($[   ]$, \infty)$

Decreasing: $($[   ]$, 4.082)$

Example 3   *Car Distance.* Two nurses, Kiara and Matias, drive away from a hospital at right angles to each other. Kiara's speed is 35 mph and Matias' is 40 mph.

a) Express the distance between the cars as a function of time, $d(t)$.

b) Find the domain of the function.

a)

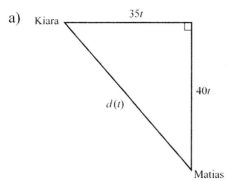

Use the Pythagorean Theorem.

$$\left[d(t)\right]^2 = (35t)^2 + (40t)^2$$
$$d(t) = \sqrt{1225t^2 + 1600t^2}$$
$$d(t) = \sqrt{2825t^2}$$
$$d(t) = [\qquad] |t|$$
$$d(t) \approx 53.15t \qquad (t \geq 0, \text{ so } |t| = t.)$$

b) The time traveled must be nonnegative, so the domain is $[$ [   ] $, \infty)$.

Example 4   A community college has 30 ft of dividers with which to set off a rectangular area for a student testing center. If a corner of the math lab is used for the testing center, the partition need only form two sides of a rectangle.

a) Express the floor area of the testing center as a function of the length of the partition.

$$A(x) = x(30 - x)$$
$$A(x) = \boxed{} - x^2$$

b) Find the domain of the function.

$$A(x) = 30x - x^2$$

The domain could be $(-\infty, \infty)$ but, since only 30 ft of dividers are available, a more realistic domain is $(0, 30)$.

c) Graph the function.

We graph the function in a window that shows the maximum value of the function. One choice that does this is $[0, 40, 0, 250]$, $Xscl = 4$, $Yscl = 25$. (Sketch the graph in the space provided below.)

d) Find the dimensions that maximize the area of the floor.

Relative maximum is 225 when $x = \boxed{}$.

Maximum area $= 225$ ft^2 when $x = 15$ ft.

Length: 15 ft

Width: $30 - \boxed{} = 15$ ft

Dimensions: 15 ft $\times$ 15 ft

Example 5   For the function defined as

$$f(x) = \begin{cases} x+1, & \text{for } x < -2, \\ 5, & \text{for } -2 \le x \le 3, \\ x^2, & \text{for } x > 3, \end{cases}$$

find $f(-5), f(-3), f(0), f(3), f(4),$ and $f(10).$

$f(-5) = -5 + 1 = \boxed{}$

$f(-3) = -3 + 1 = -2$

$f(0) = 5$

$f(3) = 5$

$f(4) = 4^2 = \boxed{}$

$f(10) = \boxed{}^2 = 100$

Example 6   Graph the function defined as

$$g(x) = \begin{cases} \dfrac{1}{3}x + 3, & \text{for } x < 3, \\ -x, & \text{for } x \ge 3. \end{cases}$$

x $(x < 3)$	$g(x) = \dfrac{1}{3}x + 3$
-3	2
0	$\boxed{}$
2	$3\dfrac{2}{3}$

x $(x \ge 3)$	$g(x) = -x$
3	$\boxed{}$
4	-4
6	$\boxed{}$

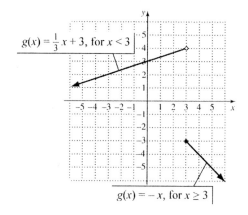

$g(x) = \dfrac{1}{3}x + 3$, for $x < 3$

$g(x) = -x$, for $x \ge 3$

Example 7 Graph the function defined as

$$f(x) = \begin{cases} 4, & \text{for } x \le 0, \\ 4 - x^2, & \text{for } 0 < x \le 2, \\ 2x - 6, & \text{for } x > 2. \end{cases}$$

x $(x \le 0)$	$f(x) = 4$
-5	4
-2	□
0	4

x $(0 < x \le 2)$	$f(x) = 4 - x^2$
$\dfrac{1}{2}$	$3\dfrac{3}{4}$
1	□
2	0

x $(x > 2)$	$f(x) = 2x - 6$
$2\dfrac{1}{2}$	-1
3	□
5	4

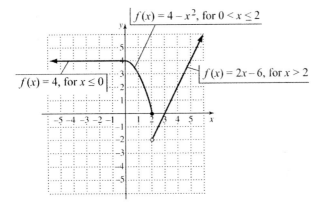

$f(x) = 4 - x^2$, for $0 < x \le 2$

$f(x) = 4$, for $x \le 0$

$f(x) = 2x - 6$, for $x > 2$

Example 8 Graph the function defined as

$$f(x) = \begin{cases} \dfrac{x^2 - 4}{x + 2}, & \text{for } x \ne -2, \\ 3, & \text{for } x = -2. \end{cases}$$

When $x \ne -2$, we have

$$\frac{x^2 - 4}{x + 2} = \frac{(x + 2)(x - 2)}{x + 2} = x - \boxed{}.$$

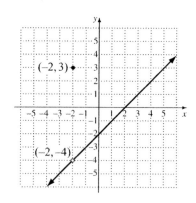

Greatest Integer Function

$f(x) = [\![x]\!]$ = the greatest integer *less than* or *equal to* x.

Example 9   Graph $f(x) = [\![x]\!]$ and determine its domain and range.

$$f(x) = \begin{cases} \vdots \\ -3, & \text{for } -3 \le x < -2, \\ -2, & \text{for } -2 \le x < -1, \\ -1, & \text{for } -1 \le x < 0, \\ 0, & \text{for } 0 \le x < 1, \\ 1, & \text{for } 1 \le x < 2, \\ -2, & \text{for } 2 \le x < 3, \\ 3, & \text{for } 3 \le x < 4, \\ \vdots \end{cases}$$

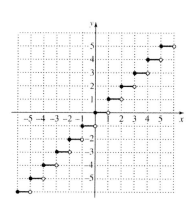

Domain: $\left(-\infty, \boxed{}\right)$

Range: $\left\{ \ldots, -3, \boxed{}, -1, 0, \boxed{}, 2, \boxed{}, \ldots \right\}$

Section 2.2 The Algebra of Functions

Sums, Differences, Products, and Quotients of Functions

If f and g are functions and x is in the domain of each function, then:

$$(f+g)(x) = f(x)+g(x),$$
$$(f-g)(x) = f(x)-g(x),$$
$$(fg)(x) = f(x) \cdot g(x),$$
$$(f/g)(x) = f(x)/g(x), \text{ provided } g(x) \neq 0.$$

Example 1   Given that $f(x) = x+1$ and $g(x) = \sqrt{x+3}$, find each of the following.

a) $(f+g)(x)$

$$(f+g)(x) = f(x)+g(x)$$
$$= x + \boxed{} + \sqrt{x+3}$$

b) $(f+g)(6)$

$$(f+g)(6) = f(6)+g\left(\boxed{}\right)$$
$$= 6+1+\sqrt{6+3}$$
$$= 7 + \sqrt{\boxed{}} = 7+3 = \boxed{}$$

We can also use the formula found in part (a) to find $(f+g)(6)$.

$$(f+g)(x) = x+1+\sqrt{x+3}$$
$$(f+g)(6) = 6+1+\sqrt{6+3}$$
$$= 7 + \sqrt{\boxed{}} = 7+3 = \boxed{}$$

c) $(f+g)(-4)$

Domain of f: $(-\infty, \infty)$

Domain of g: $[-3, \infty)$ because $x+3 \geq 0$ when $x \geq \boxed{}$

-4 is not in the domain of g so $(f+g)(-4)$ does not exist.

> **Domains of $f + g, f - g, fg,$ and f / g**
>
> If f and g are functions, then the domain of the functions $f + g, f - g,$ and fg is the intersection of the domain of f and the domain of g. The domain of f / g is also the intersection of the domains of f and g with the exclusion of any x-values for which $g(x) = 0$.

Example 2   Given that $f(x) = x^2 - 4$ and $g(x) = x + 2$, find each of the following.

a) The domain of $f + g$, $f - g$, fg, and f / g.

   The domain of f is the set of all real numbers, and the domain of g is also the set of all real numbers.

   Domain of $f + g$: $(-\infty, \infty)$

   Domain of $f - g$: $(-\infty, \infty)$

   Domain of fg: $(-\infty, \infty)$

   $f / g = \dfrac{x^2 - 4}{x + 2}$ is not defined when $x = -2$, so the domain of f / g is

   $(-\infty, -2) \cup \left(\boxed{}, \infty \right)$.

b) $(f + g)(x)$

   $(f + g)(x) = f(x) + g(x)$

   $\qquad = x^2 - 4 + x + 2$

   $\qquad = x^2 + \boxed{} - 2$

c) $(f - g)(x)$

   $(f - g)(x) = f(x) - g(x)$

   $\qquad = x^2 - 4 - \left(x + \boxed{} \right)$

   $\qquad = x^2 - 4 - x - 2$

   $\qquad = x^2 - x - \boxed{}$

d) $(fg)(x)$

   $(fg)(x) = f(x) \cdot g(x)$

   $\qquad = \left(x^2 - \boxed{} \right)(x + 2)$

   $\qquad = x^3 + 2x^2 - 4x - \boxed{}$

e) $(f/g)(x)$

$$(f/g)(x) = \frac{f(x)}{g(x)} = \frac{x^2 - 4}{x + 2}$$

$$= \frac{\left(x + \boxed{}\right)(x - 2)}{x + 2}$$

$$= x - \boxed{}, \text{ where } x \neq -2$$

f) $(gg)(x)$

$$(gg)(x) = g(x) \cdot g(x) = \left[g(x)\right]^2$$

$$= \left(x + \boxed{}\right)^2$$

$$= x^2 + 4x + \boxed{}$$

Example 3   For the function f given by $f(x) = 2x - 3$, find and simplify the difference quotient

$$\frac{f(x + h) - f(x)}{h}.$$

$$\frac{f(x + h) - f(x)}{h} = \frac{2(x + h) - \boxed{} - (2x - 3)}{h}$$

$$= \frac{2x + 2h - 3 - 2x + 3}{h}$$

$$= \frac{2h}{\boxed{}}$$

$$= 2$$

Example 4   For the function f given by $f(x) = \dfrac{1}{x}$, find and simplify the difference quotient

$$\frac{f(x+h) - f(x)}{h}.$$

$$\frac{f(x+h) - f(x)}{h} = \frac{\dfrac{1}{x+h} - \dfrac{1}{x}}{h}$$

$$= \frac{\dfrac{1}{x+h} \cdot \dfrac{x}{x} - \dfrac{1}{x} \cdot \dfrac{x+h}{x+h}}{\boxed{}}$$

$$= \frac{\dfrac{x}{x(x+h)} - \dfrac{x+h}{x(x+h)}}{h}$$

$$= \frac{\dfrac{x - (x+h)}{x(x+h)}}{h} = \frac{\dfrac{x - x - \boxed{}}{x(x+h)}}{h}$$

$$= \frac{\dfrac{-h}{x(x+h)}}{h} = \frac{-h}{x(x+h)} \cdot \frac{\boxed{}}{h}$$

$$= \frac{-h \cdot 1}{x \cdot (x+h) \cdot h}$$

$$= \frac{\boxed{}}{x(x+h)}, \text{ or } -\frac{1}{x(x+h)}$$

Example 5   For the function f given by $f(x) = 2x^2 - x - 3$, find and simplify the difference quotient

$$\frac{f(x+h) - f(x)}{h}.$$

We first find $f(x+h)$.

$$f(x+h) = 2(x+h)^2 - (x+h) - 3$$
$$= 2(x^2 + 2xh + h^2) - \left(x + \boxed{}\right) - 3$$
$$= 2x^2 + 4xh + 2h^2 - x - h - \boxed{}$$

Then we have

$$\frac{f(x+h) - f(x)}{h}$$

$$= \frac{\left(2x^2 + 4xh + 2h^2 - x - \boxed{} - 3\right) - \left(2x^2 - x - 3\right)}{h}$$

$$= \frac{2x^2 + 4xh + 2h^2 - \boxed{} - h - 3 - 2x^2 + \boxed{} + 3}{h}$$

$$= \frac{4xh + 2h^2 - h}{h} = \frac{h\left(4x + 2h - \boxed{}\right)}{h \cdot 1}$$

$$= \boxed{} x + 2h - 1$$

Section 2.3   The Composition of Functions

Composition of Functions

The **composite function** $f \circ g$, the **composition** of f and g, is defined as

$$(f \circ g)(x) = f(g(x)),$$

where x is in the domain of g and $g(x)$ is in the domain of f.

Example 1   Given that $f(x) = 2x - 5$ and $g(x) = x^2 - 3x + 8$, find each of the following.

a) $(f \circ g)(x)$ and $(g \circ f)(x)$

$$(f \circ g)(x) = f(g(x)) = f\left(x^2 - 3x + \boxed{}\right)$$
$$= 2\left(x^2 - 3x + 8\right) - 5 = 2x^2 - 6x + \boxed{} - 5$$
$$= 2x^2 - 6x + 11$$

$$(g \circ f)(x) = g(f(x)) = g\left(2x - \boxed{}\right)$$
$$= (2x - 5)^2 - 3(2x - 5) + 8$$
$$= 4x^2 - \boxed{}x + 25 - 6x + 15 + 8$$
$$= 4x^2 - 26x + \boxed{}$$

b) $(f \circ g)(7)$ and $(g \circ f)(7)$

$$(f \circ g)(7) = f\left(7^2 - 3 \cdot \boxed{} + 8\right) = f(36)$$
$$= 2 \cdot \boxed{} - 5 = 67$$

$$(g \circ f)(7) = g(2 \cdot 7 - 5) = g\left(\boxed{}\right)$$
$$= 9^2 - 3 \cdot 9 + \boxed{} = 62$$

We could also use the results from part (a) to find $(f \circ g)(7)$ and $(g \circ f)(7)$.

$$(f \circ g)(x) = 2x^2 - 6x + 11$$
$$(f \circ g)(7) = 2 \cdot 7^2 - 6 \cdot \boxed{} + 11 = 67$$

$$(g \circ f)(x) = 4x^2 - 26x + 48$$
$$(g \circ f)(7) = 4 \cdot 7^2 - 26 \cdot 7 + \boxed{} = 62$$

c) $(g \circ g)(1)$

$$(g \circ g)(1) = g(g(1)) = g\left(1^2 - 3 \cdot \boxed{} + 8\right)$$

$$= g(6)$$

$$= \boxed{}^2 - 3 \cdot 6 + 8 = 26$$

d) $(f \circ f)(x)$

$$(f \circ f)(x) = f(f(x)) = f(2x - 5)$$

$$= 2\left(2x - \boxed{}\right) - 5$$

$$= \boxed{} x - 10 - 5$$

$$= 4x - 15$$

Example 2

a) Given that $f(x) = \sqrt{x}$ and $g(x) = x - 3$, find $f \circ g$ and $g \circ f$.

$$(f \circ g)(x) = f(g(x)) = \sqrt{x - 3}$$

$$(g \circ f)(x) = g(f(x)) = \boxed{} - 3$$

b) Given that $f(x) = \sqrt{x}$ and $g(x) = x - 3$, find the domain of $f \circ g$ and the domain of $g \circ f$.

$$(f \circ g)(x) = \sqrt{x - 3}$$

Domain of $f(x) = \left[\boxed{}, \infty\right)$

Domain of $g(x) = (-\infty, \infty)$

Domain of $f \circ g = \{x | x \geq 3\}$, or $\left[\boxed{}, \infty\right)$ because $g(x) \geq 0$, or $x - 3 \geq 0$, when $x \geq 3$.

We could also get this by considering the domain of $(f \circ g)(x) = \sqrt{x - 3}$.

$$(g \circ f)(x) = \sqrt{x} - 3$$

Domain of $g \circ f =$ Domain of $f(x) = [0, \infty)$

We could also find this by considering the domain of $(g \circ f)(x) = \sqrt{x} - 3$.

Example 3   Given that $f(x) = \dfrac{1}{x-2}$ and $g(x) = \dfrac{5}{x}$, find $f \circ g$ and $g \circ f$ and the domain of each.

$$(f \circ g)(x) = f(g(x)) = \dfrac{1}{\dfrac{5}{x} - 2}$$

$$= \dfrac{1}{\dfrac{5 - 2x}{x}} = 1 \cdot \dfrac{x}{5 - 2x}$$

$$= \dfrac{x}{5 - 2x}$$

$$(g \circ f)(x) = g(f(x)) = \dfrac{5}{\dfrac{1}{x-2}}$$

$$= 5 \cdot \dfrac{x - 2}{1}$$

$$= 5(x - 2)$$

Domain of g: $\{x \mid x \neq 0\}$

Domain of f: $\{x \mid x \neq 2\}$

In addition to 0, values of x for which $g(x) = 2$ must be excluded from the domain of $f \circ g$. We have

$$\dfrac{5}{x} = 2$$

$$5 = 2x$$

$$\dfrac{5}{2} = x.$$

Both $\dfrac{5}{2}$ and $\boxed{}$ must be excluded from the domain of $f \circ g$.

Domain of $f \circ g$: $\left(-\infty, \boxed{}\right) \cup \left(0, \dfrac{5}{2}\right) \cup \left(\boxed{}, \infty\right)$

In addition to 2, values of x for which $f(x) = 0$ must be excluded from the domain of $g \circ f$. We have

$$\frac{1}{x-2} = 0$$

$$\boxed{} = 0 \qquad \text{False}$$

Thus, only 2 must be excluded.

Domain of $g \circ f$: $\left(-\infty, \boxed{}\right) \cup (2, \infty)$

Example 4   If $h(x) = (2x - 3)^5$, find $f(x)$ and $g(x)$ such that $h(x) = (f \circ g)(x)$.

For $f(x) = x^5$ and $g(x) = 2x - \boxed{}$,

$$h(x) = (f \circ g)(x) = f(g(x))$$
$$= f(2x - 3)$$
$$= \left(\boxed{}\right)^5.$$

These functions give us $h(x)$, but there are other choices for $f(x)$ and $g(x)$.

For $f(x) = (x + 7)^5$ and $g(x) = 2x - 10$,

$$h(x) = (f \circ g)(x) = f(g(x))$$
$$= f\left(2x - \boxed{}\right)$$
$$= (2x - 10 + 7)^5$$
$$= \left(2x - \boxed{}\right)^5.$$

This is also a correct choice. There are others as well.

Example 5   If $h(x) = \dfrac{1}{(x+3)^3}$, find $f(x)$ and $g(x)$ such that $h(x) = (f \circ g)(x)$.

For $f(x) = \dfrac{1}{x^3}$ and $g(x) = x + \boxed{}$,

$$h(x) = (f \circ g)(x) = f(g(x))$$
$$= f(\boxed{})$$
$$= \frac{1}{(x+3)^3}.$$

These functions give us $h(x)$, but there are other correct choices for $f(x)$ and $g(x)$.

For $f(x) = \dfrac{1}{x}$ and $g(x) = (x+3)^3$,

$$h(x) = (f \circ g)(x) = f(g(x))$$
$$= f\left((x+3)^3\right)$$
$$= \frac{1}{\boxed{}}.$$

This is also a correct choice. There are others as well.

Section 2.4 Symmetry

Algebraic Tests of Symmetry

x-axis: If replacing y with $-y$ produces an equivalent equation, then the graph is *symmetric with respect to the x-axis.*

y-axis: If replacing x with $-x$ produces an equivalent equation, then the graph is *symmetric with respect to the y-axis.*

Origin: If replacing x with $-x$ and y with $-y$ produces an equivalent equation, then the graph is *symmetric with respect to the origin.*

Example 1   Test $y = x^2 + 2$ for symmetry with respect to the x-axis, the y-axis, and the origin.

x-axis:

$y = x^2 + 2$

$\downarrow$

$-y = x^2 + 2$ Replacing y with $-y$

$y = -x^2 - \boxed{}$ Not equivalent to $y = x^2 + 2$

The graph $\underset{\text{is / is not}}{\underline{}}$ symmetric with respect to the x-axis.

y-axis:

$y = x^2 + 2$

$\downarrow$

$y = (-x)^2 + 2$ Replacing x with $-x$

$y = \boxed{} + 2$ Obtaining the original equation

The graph $\underset{\text{is / is not}}{\underline{}}$ symmetric with respect to the y-axis.

Origin:

$y = x^2 + 2$

$\downarrow \quad \downarrow$

$-y = (-x)^2 + 2$ Replacing y with $-y$ and x with $-x$

$\boxed{} = x^2 + 2$

$\boxed{} = -x^2 - 2$ Not equivalent to $y = x^2 + 2$

The graph $\underset{\text{is / is not}}{\underline{}}$ symmetric with respect to the origin.

Example 2   Test $x^2 + y^4 = 5$ for symmetry with respect to the x-axis, the y-axis, and the origin.

x-axis:

$$x^2 + y^4 = 5$$
$$\downarrow$$
$$x^2 + (-y)^4 = 5 \qquad \text{Replacing } y \text{ with } -y$$
$$x^2 + \boxed{} = 5 \qquad \text{Obtaining the original equation}$$

The graph _____ symmetric with respect to the x-axis.
$\qquad\qquad$ is / is not

y-axis:

$$x^2 + y^4 = 5$$
$$\downarrow$$
$$(-x)^2 + y^4 = 5 \qquad \text{Replacing } x \text{ with } -x$$
$$x^2 + y^4 = \boxed{} \qquad \text{Obtaining the original equation}$$

The graph _____ symmetric with respect to the y-axis.
$\qquad\qquad$ is / is not

Origin:

$$x^2 + y^4 = 5$$
$$\downarrow \qquad \downarrow$$
$$(-x)^2 + (-y)^4 = 5 \qquad \text{Replacing } x \text{ with } -x \text{ and } y \text{ with } -y$$
$$\boxed{} + y^4 = 5 \qquad \text{Obtaining the original equation}$$

The graph _____ symmetric with respect to the origin.
$\qquad\qquad$ is / is not

Even Functions and Odd Functions

If the graph of a function f is symmetric with respect to the y-axis, we say that it is an **even function**. That is, for each x in the domain of f, $f(x) = f(-x)$.

If the graph of a function f is symmetric with respect to the origin, we say that it is an **odd function**. That is, for each x in the domain of f, $f(-x) = -f(x)$.

Example 3

a) Determine whether $f(x) = 5x^7 - 6x^3 - 2x$ is odd, even, or neither.

$$f(-x) = 5(-x)^7 - 6(-x)^3 - 2(-x)$$
$$= -5x^7 + 6x^3 + 2x$$

$$f(-x) \neq f(x)$$

Thus, $f(x)$ _____ even.
is / is not

$$-f(x) = -\left(5x^7 - 6x^3 - 2x\right)$$
$$= -5x^7 + 6x^3 + 2x$$

$$f(-x) = -f(x)$$

Thus, $f(x)$ _____ odd.
is / is not

b) Determine if $h(x) = 5x^6 - 3x^2 - 7$ is even, odd, or neither.

$$h(-x) = 5(-x)^6 - 3(-x)^2 - 7$$
$$= 5x^6 - 3x^2 - \boxed{}$$

$$h(-x) = h(x)$$

Thus, $h(x)$ is an _____ function.
even / odd

The function cannot be both even and odd, so we stop here.

Section 2.5 Transformations

> **Vertical Translation and Horizontal Translation**
>
> For $b > 0$:
>
>   the graph of $y = f(x) + b$ is the graph of $y = f(x)$ shifted *up* b units;
>
>   the graph of $y = f(x) - b$ is the graph of $y = f(x)$ shifted *down* b units.
>
> For $d > 0$:
>
>   the graph of $y = f(x - d)$ is the graph of $y = f(x)$ shifted *to the right* d units;
>
>   the graph of $y = f(x + d)$ is the graph of $y = f(x)$ shifted *to the left* d units.

Example 1   Graph each of the following. Before doing so, describe how each graph can be obtained from one of the basic graphs shown earlier in this section.

a) $g(x) = x^2 - 6$

We compare this with the graph of $f(x) = x^2$.

$g(x) = f(x) - 6$, so the graph of $g(x)$ is the graph of $f(x)$ translated

_____ 6 units.
  up / down

We compare some points on the graphs of *f* and *g*.

$$\underline{f} \qquad \underline{g}$$

$(-3, 9) \quad (-3, 3)$

$(0, 0) \quad (0, \boxed{})$

$(2, 4) \quad (2, \boxed{})$

The *y*-coordinate of a point on the graph of $g(x)$ is 6 _____ than the
                                               more / less

corresponding point on the graph of $f(x)$.

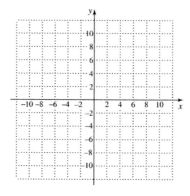

b) $g(x) = |x - 4|$

We compare this with the graph of $f(x) = |x|$.

$g(x) = f(x - 4)$, so the graph of $g(x)$ is the graph of $f(x)$ shifted

_____ 4 units.
right / left

We compare some points on the graphs of f and g.

f	g
$(-4, 4)$	$(0, 4)$
$(0, 0)$	$(\boxed{}, 0)$
$(6, 6)$	$(10, \boxed{})$

The x-coordinate of a point on the graph of $g(x)$ is 4 _____ than the
more / less

corresponding point on the graph of $f(x)$.

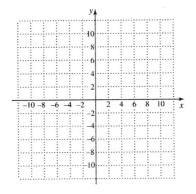

c) $g(x) = \sqrt{x + 2}$

We compare this with the graph of $f(x) = \sqrt{x}$.

$g(x) = f(x + 2)$, so the graph of $g(x)$ is the graph of $f(x)$ shifted

_____ 2 units, as we can see from the graphs of the functions.
right / left

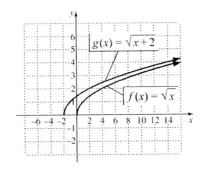

d) $h(x) = \sqrt{x+2} - 3$

We compare this with the graph of $f(x) = \sqrt{x}$.

In part (c) we saw that the graph of $g(x) = \sqrt{x+2}$ is the graph of $f(x)$ shifted left ☐ units.

$h(x) = g(x) - 3$, so the graph of $h(x)$ is the graph of $g(x)$ shifted _____ 3 units.

up / down

Together, the graph of $h(x)$ is the graph of $f(x) = \sqrt{x}$ shifted left ☐ units and down ☐ units.

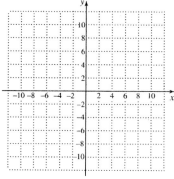

Reflections

The graph of $y = -f(x)$ is the **reflection** of the graph of $y = f(x)$ across the x-axis.

The graph of $y = f(-x)$ is the **reflection** of the graph of $y = f(x)$ across the y-axis.

If a point (x, y) is on the graph of $y = f(x)$, then $(x, -y)$ is on the graph of $y = -f(x)$, and $(-x, y)$ is on the graph of $y = f(-x)$.

Example 2   Graph each of the following. Before doing so, describe how each graph can be obtained from the graph of $f(x) = x^3 - 4x^2$.

a) $g(x) = (-x)^3 - 4(-x)^2$

$f(-x) = (-x)^3 - 4(-x)^2 = g(x)$, so $g(x)$ is a reflection of $f(x)$ across the

_____ axis.

x- / y-

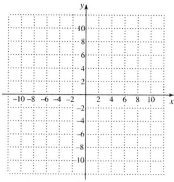

If (x, y) is on the graph of f, then $\left(\boxed{}, y\right)$ is on the graph of g. For example, if $(2, -8)$ is on the graph of f, then $(-2, -8)$ is on the graph of g.

b) $h(x) = 4x^2 - x^3$

$-f(x) = -(x^3 - 4x^2) = -x^3 + 4x^2 = 4x^2 - x^3 = h(x)$, so the graph of $h(x)$ is the

reflection of the graph of $f(x)$ across the _____ axis.

x- / y-

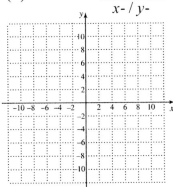

If (x, y) is on the graph of f, then $\left(x, \boxed{}\right)$ is on the graph of h. For example, if $(2, -8)$

is on the graph of f, then $(2, 8)$ is on the graph of h.

Vertical Stretching and Shrinking and Horizontal Stretching and Shrinking

The graph of $y = af(x)$ can be obtained from the graph of $y = f(x)$ by

   stretching vertically for $|a| > 1$, or

   shrinking vertically for $0 < |a| < 1$.

For $a < 0$, the graph is also reflected across the x-axis. (The y-coordinates of the graph of $y = af(x)$ can be obtained by multiplying the y-coordinates of $y = f(x)$ by a.)

The graph of $y = f(cx)$ can be obtained from the graph of $y = f(x)$ by

   shrinking horizontally for $|c| > 1$, or

   stretching horizontally for $0 < |c| < 1$.

For $c < 0$, the graph is also reflected across the y-axis. (The x-coordinates of the graph of $y = f(cx)$ can be obtained by dividing the x-coordinates of the graph of $y = f(x)$ by c.)

Example 3 Shown is a graph of $y = f(x)$ for some function f. No formula for f is given.

Graph each of the following.

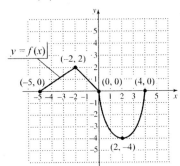

a) $g(x) = 2f(x)$

The graph of $y = af(x)$ is a vertical stretching or shrinking of the graph of $y = f(x)$. Here $|a| = |2| > 1$, so this is a stretching. Each y-value on $f(x)$ will be multiplied by 2 to obtain the graph of $g(x)$.

Points on f	Corresponding points on g
$(-5, 0)$	$(-5, 0)$
$(-2, 2)$	$\left(-2, \boxed{}\right)$
$(0, 0)$	$(0, 0)$
$(2, -4)$	$\left(2, \boxed{}\right)$
$(4, 0)$	$(4, 0)$

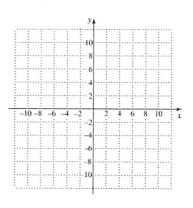

b) $h(x) = \dfrac{1}{2} f(x)$

The graph of $y = af(x)$ is a vertical stretching or shrinking of the graph of $y = f(x)$. Here $|a| = \left|\dfrac{1}{2}\right| < 1$, so this is a shrinking. Each y-value on $f(x)$ will be multiplied by $\dfrac{1}{2}$ to obtain the graph of $h(x)$.

Points on f	Corresponding points on h
$(-5, 0)$	$\left(-5, \boxed{}\right)$
$(-2, 2)$	$(-2, 1)$
$(0, 0)$	$\left(0, \boxed{}\right)$
$(2, -4)$	$\left(2, \boxed{}\right)$
$(4, 0)$	$(4, 0)$

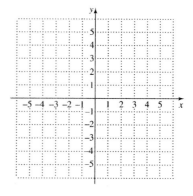

c) $r(x) = f(2x)$

The graph of $y = f(x)$ is a horizontal stretching or shrinking. Here $|c| = |2| > 1$, so this is a shrinking. Each x-coordinate on $f(x)$ will be divided by 2 to obtain the graph of r.

Points on f	Corresponding points on r
$(-5, 0)$	$(-2.5, 0)$
$(-2, 2)$	$(\boxed{}, 2)$
$(0, 0)$	$(\boxed{}, 0)$
$(2, -4)$	$(1, -4)$
$(4, 0)$	$(\boxed{}, 0)$

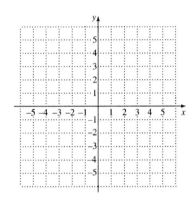

d) $s(x) = f\left(\dfrac{1}{2}x\right)$

The graph of $y = f(x)$ is a horizontal stretching or shrinking. Here $|c| = \left|\dfrac{1}{2}\right| < 1$, so this is a stretching. Each x-coordinate on $f(x)$ will be multiplied by 2 to obtain the graph of s.

Points on f	Corresponding points on s
$(-5, 0)$	$(\boxed{}, 0)$
$(-2, 2)$	$(-4, 2)$
$(0, 0)$	$(\boxed{}, 0)$
$(2, -4)$	$(\boxed{}, -4)$
$(4, 0)$	$(8, 0)$

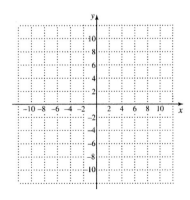

e) $t(x) = f\left(-\dfrac{1}{2}x\right)$

The graph of $y = f(x)$ is a horizontal stretching or shrinking. Here $c = -\dfrac{1}{2}$. Because $c < 0$, we will also have the reflection of f across the y-axis. We have

$|c| = \left|-\dfrac{1}{2}\right| = \boxed{}$ <1, so this is a horizontal stretching along with the reflection across the y-axis. We will multiply each x-coordinate of f by 2 and then change its sign.

Points on f	Corresponding points on t
$(-5, 0)$	$(10, 0)$
$(-2, 2)$	$\left(\boxed{}, 2\right)$
$(0, 0)$	$(0, 0)$
$(2, -4)$	$\left(\boxed{}, -4\right)$
$(4, 0)$	$\left(\boxed{}, 0\right)$

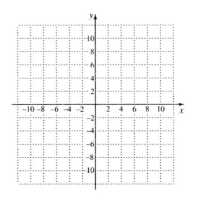

Example 4   Use the graph of $y = f(x)$ shown below to graph $y = -2f(x-3) + 1$.

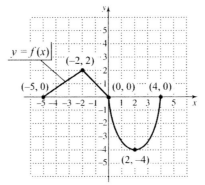

First consider $y = f(x-3)$. This is a shift of the given graph $\boxed{}$ units to the $\underline{\hspace{1.5cm}}$.
 right / left

Points on $y = f(x)$	Corresponding points on $y = f(x-3)$
$(-5, 0)$	$(-2, 0)$
$(-2, 2)$	$\left(\boxed{}, 2\right)$
$(0, 0)$	$(3, 0)$
$(2, -4)$	$\left(\boxed{}, -4\right)$
$(4, 0)$	$(7, 0)$

Now sketch the graph of $y = f(x-3)$.

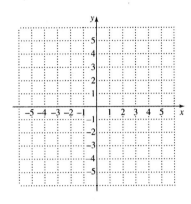

Next consider $y = 2f(x-3)$. This will stretch the graph of $y = f(x-3)$ vertically by a factor of 2.

Points on $y = f(x-3)$	Corresponding points on $y = 2f(x-3)$
$(-2, 0)$	$(-2, 0)$
$(3, 0)$	$(3, 0)$
$(7, 0)$	$\left(7, \boxed{}\right)$
$(1, 2)$	$\left(1, \boxed{}\right)$
$(5, -4)$	$(5, -8)$

Now sketch the graph of $y = 2f(x-3)$.

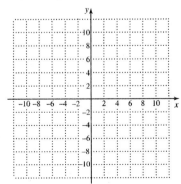

Next consider $y = -2f(x-3)$. This is the reflection of the graph of $y = 2f(x-3)$ across the _____ axis.
        x- / y-

Points on $y = 2f(x-3)$	Corresponding points on $y = -2f(x-3)$
$(-2, 0)$	$(-2, 0)$
$(3, 0)$	$\left(3, \boxed{}\right)$
$(7, 0)$	$(7, 0)$
$(1, 4)$	$\left(1, \boxed{}\right)$
$(5, -8)$	$\left(5, \boxed{}\right)$

Now sketch the graph of $y = -2f(x-3)$.

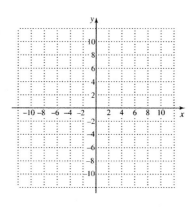

Finally, consider $y = -2f(x-3)+1$. This is a shift of the graph of $y = -2f(x-3)$

_____ 1 unit.
up / down

Points on $y = -2f(x-3)$	Corresponding points on $y = -2f(x-3)+1$
$(-2, 0)$	$(-2, 1)$
$(1, -4)$	$\left(1, \boxed{}\right)$
$(3, 0)$	$\left(3, \boxed{}\right)$
$(5, 8)$	$(5, 9)$
$(7, 0)$	$\left(7, \boxed{}\right)$

Now we can sketch the graph of $y = -2f(x-3)+1$.

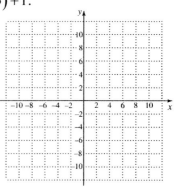

Section 2.6   Variation and Applications

Direct Variation

If a situation gives rise to a linear function $f(x) = kx$, or $y = kx$, where k is a positive constant, we say that we have **direct variation**, or that y **varies directly as x**, or that y **is directly proportional to** x. The number k is called the **variation constant**, or the **constant of proportionality.**

Example 1   Find the variation constant and an equation of variation in which y varies directly as x, and $y = 32$ when $x = 2$.

$$y = kx$$

$$32 = k \cdot \boxed{}$$

$$\boxed{} = k$$

The variation constant is 16.

The equation of variation is $y = \boxed{} x.$

Example 2   *Water from Melting Snow.* The number of centimeters W of water produced from melting snow varies directly as S, the number of centimeters of snow. Meteorologists have found that under certain conditions 150 cm of snow will melt to 16.8 cm of water. To how many centimeters of water will 200 cm of snow melt?

$$y = kx$$

$$W(S) = kS \qquad \text{Using function notation}$$

$$16.8 = k \cdot \boxed{}$$

$$0.112 = \boxed{} \qquad \text{Variation constant}$$

$$W(S) = 0.112S \qquad \text{Equation of variation}$$

$$W(200) = 0.112 \left(\boxed{} \right)$$

$$= 22.4$$

Thus, 200 cm of snow will melt to $\boxed{}$ cm of water.

Inverse Variation

If a situation gives rise to a function $f(x) = k/x$, or $y = k/x$, where k is a positive constant, we say that we have **inverse variation**, or that y **varies inversely as** x, or that y **is inversely proportional to** x. The number k is called the **variation constant**, or the **constant of proportionality**.

Example 3   Find the variation constant and an equation of variation in which y varies inversely as x, and $y = 16$ when $x = 0.3$.

$$y = \frac{k}{x}$$

$$\boxed{} = \frac{k}{0.3}$$

$$\boxed{} = k$$

The variation constant is $\boxed{}$.

The equation of variation is $y = \dfrac{\boxed{}}{x}$.

Example 4   The time t required to fill a swimming pool varies inversely as the rate of flow r of water into the pool. A tank truck can fill a pool in 90 min at a rate of 1500 L/min. How long would it take to fill the pool at a rate of 1800 L/min?

$$t(r) = \frac{k}{r}$$

$$t(1500) = \frac{k}{1500} \qquad \text{Substituting 1500 for } r$$

$$90 = \frac{k}{\boxed{}} \qquad \text{Replacing } t(1500) \text{ with 90}$$

$$90 \cdot 1500 = k$$

$$\boxed{} = k \qquad \text{Variation constant}$$

$$t(r) = \frac{135,000}{r} \qquad \text{Equation of variation}$$

$$t(1800) = \frac{135,000}{\boxed{}}$$

$$t(1800) = \boxed{}$$

It would take $\boxed{}$ min to fill the pool at a rate of 1800 L/min.

---

y varies **directly as the nth power of x** if there is some positive constant k such that

$$y = kx^n.$$

y varies **inversely as the nth power of x** if there is some positive constant k such that

$$y = \frac{k}{x^n}.$$

y varies **jointly as x and z** if there is some positive constant k such that

$$y = kxz.$$

Example 5   Find an equation of variation in which y varies directly as the square of x, and $y = 12$ when $x = 2$.

$$y = kx^2$$
$$\boxed{} = k \cdot 2^2$$
$$12 = k \cdot 4$$
$$\boxed{} = k$$

The equation of variation is $y = \boxed{} x^2$.

Example 6   Find an equation of variation in which y varies jointly as x and z, and $y = 42$ when $x = 2$ and $z = 3$.

$$y = kxz$$
$$\boxed{} = k \cdot 2 \cdot 3$$
$$42 = k \cdot \boxed{}$$
$$\boxed{} = k$$

The equation of variation is $y = \boxed{} xz$.

Example 7   Find an equation of variation in which y varies jointly as x and z and inversely as the square of w, and $y = 105$ when $x = 3$, $z = 20$, and $w = 2$.

$$y = k \cdot \frac{xz}{w^2}$$

$$105 = \frac{k \cdot \boxed{} \cdot 20}{2^2}$$

$$105 = \frac{k \cdot 60}{\boxed{}}$$

$$105 = k \cdot 15$$

$$\boxed{} = k$$

The equation of variation is $y = \boxed{} \cdot \dfrac{xz}{w^2} = \dfrac{7xz}{w^2}$.

Example 8   *Volume of a Tree.* The volume of wood V in a tree varies jointly as the height h and the square of the girth g. (Girth is distance around.) If the volume of a redwood tree is 216 m³ when the height is 30 m and the girth is 1.5 m, what is the height of a tree whose volume is 344 m³ and whose girth is 1.6 m?

$$V = k \cdot h \cdot g^2$$

$$216 = k \cdot \boxed{} \cdot (1.5)^2$$

$$216 = k \cdot 30 \cdot 2.25$$

$$216 = k \cdot 67.5$$

$$\boxed{} = k \qquad\qquad \text{Variation constant}$$

$$V = 3.2 h g^2 \qquad\qquad \text{Equation of variation}$$

$$\boxed{} = 3.2 \cdot h \cdot 1.6^2$$

$$344 = 8.192 h$$

$$\boxed{} \approx h$$

The height is $\boxed{}$ m.

Section 3.1 The Complex Numbers

> **The Number i**
>
> The number i is defined such that
>
> $$i = \sqrt{-1} \text{ and } i^2 = -1.$$

Example 1 Express each number in terms of i.

a) $\sqrt{-7} = \sqrt{-1 \cdot 7}$

$\qquad = \boxed{} \cdot \sqrt{7}$

$\qquad = \boxed{} \sqrt{7}$, or $\sqrt{7}\,i$

b) $\sqrt{-16} = \sqrt{-1 \cdot 16}$

$\qquad = \boxed{} \cdot \sqrt{16}$

$\qquad = i \cdot \boxed{}$

$\qquad = 4i$

c) $-\sqrt{-13} = -\sqrt{-1 \cdot 13}$

$\qquad = -\sqrt{-1} \cdot \sqrt{13}$

$\qquad = -\boxed{} \sqrt{13}$, or $-\sqrt{13}\,i$

d) $-\sqrt{-64} = -\sqrt{-1 \cdot 64}$

$\qquad = -\boxed{} \cdot \sqrt{64}$

$\qquad = -i \cdot \boxed{}$

$\qquad = -8i$

e) $\sqrt{-48} = \sqrt{-1 \cdot 48}$

$\qquad = \sqrt{-1} \cdot \sqrt{48}$

$\qquad = \boxed{} \cdot \sqrt{48}$

$\qquad = i \cdot \sqrt{16 \cdot 3}$

$\qquad = i \cdot \boxed{} \cdot \sqrt{3}$

$\qquad = i \cdot 4 \cdot \sqrt{3}$

$\qquad = \boxed{}$

> **Complex Numbers**
>
> A **complex number** is a number of the form $a + bi$, where a and b are real numbers. The number a is said to be the **real part** of $a + bi$ and the number b is said to be the **imaginary part** of $a + bi$.

Example 2   Add or subtract and simplify each of the following.

a) $(8 + 6i) + (3 + 2i)$

$$(8 + 6i) + (3 + 2i) = \left(8 + \boxed{}\right) + \left(6i + \boxed{}\right)$$

$$= 11 + \boxed{}$$

b) $(4 + 5i) - (6 - 3i)$

$$(4 + 5i) - (6 - 3i) = (4 - 6) + \left(5i + \boxed{}\right)$$

$$= \boxed{} + 8i$$

Example 3   Multiply and simplify each of the following.

a) $\sqrt{-16} \cdot \sqrt{-25} = \boxed{} \cdot \sqrt{16} \cdot \boxed{} \cdot \sqrt{25}$

$$= \boxed{} \cdot 4 \cdot i \cdot \boxed{}$$

$$= 20i^2 \qquad\qquad i = \sqrt{-1}$$

$$= 20\left(\boxed{}\right) \qquad\qquad i^2 = -1$$

$$= -20$$

b) $(1 + 2i)(1 + 3i) = 1 + 3i + 2i + \boxed{}$

$$= 1 + 3i + 2i + 6\left(\boxed{}\right)$$

$$= 1 + \boxed{} - 6$$

$$= -5 + 5i$$

c) $(3 - 7i)^2 = 9 - \boxed{} + 49i^2$

$$= 9 - 42i + 49(-1)$$

$$= 9 - 42i - 49$$

$$= \boxed{} - 42i$$

Example 4 Simplify each of the following.

a) $i^{37} = \boxed{} \cdot i$

$= \left(i^2\right)^{\boxed{}} \cdot i$

$= \left(\boxed{}\right)^{18} \cdot i$

$= 1 \cdot i$

$= i$

b) $i^{58} = \left(i^2\right)^{\boxed{}}$

$= \left(-1\right)^{29}$

$= \boxed{}$

c) $i^{75} = \boxed{} \cdot i$

$= \left(i^2\right)^{\boxed{}} \cdot i$

$= \left(-1\right)^{37} \cdot i$

$= -1 \cdot i$

$= \boxed{}$

d) $i^{80} = \left(i^2\right)^{\boxed{}}$

$= \left(-1\right)^{40}$

$= \boxed{}$

Conjugate of a Complex Number

The **conjugate** of a complex number $a + bi$ is $a - bi$. The numbers $a + bi$ and $a - bi$ are **complex conjugates**.

Example 5 Multiply each of the following.

a) $(5 + 7i)(5 - 7i)$ $\qquad (A + B)(A - B) = A^2 - B^2$

$= 5^2 - \left(\boxed{}\right)^2$

$= 25 - 49i^2$

$= 25 - 49\left(\boxed{}\right)$

$= 25 + 49$

$= \boxed{}$

b) $(8i)(-8i) = -64i^2$

$= -64\left(\boxed{}\right)$

$= 64$

Example 6   Divide $2-5i$ by $1-6i$.

$$\frac{2-5i}{1-6i} = \frac{2-5i}{1-6i} \cdot \frac{\boxed{}}{\boxed{}} = \frac{(2-5i)(1+6i)}{(1-6i)(1+6i)}$$

$$= \frac{2+12i-5i-\boxed{}}{1-\left(\boxed{}\right)^2}$$

$$= \frac{2+12i-5i+\boxed{}}{1-36i^2} = \frac{\boxed{}+7i}{1+36}$$

$$= \frac{32+7i}{\boxed{}} = \frac{32}{37}+\frac{7}{37}i$$

Section 3.2   Quadratic Equations, Functions, Zeros, and Models

Quadratic Equations

A **quadratic equation** is an equation that can be written in the form

$ax^2 + bx + c = 0$, $a \neq 0$, where a, b, and c are real numbers.

Quadratic Functions

A **quadratic function** f is a function that can be written in the form

$f(x) = ax^2 + bx + c$, $a \neq 0$, where a, b, and c are real numbers.

The **zeros** of a quadratic function $f(x) = ax^2 + bx + c$ are the *solutions* of the associated quadratic equation $ax^2 + bx + c = 0$.

Equation-Solving Principles

The Principle of Zero Products: If $ab = 0$ is true, then $a = 0$ or $b = 0$, and if $a = 0$ or $b = 0$, then $ab = 0$.

The Principle of Square Roots: If $x^2 = k$, then $x = \sqrt{k}$ or $x = -\sqrt{k}$.

Example 1   Solve: $2x^2 - x = 3$.

$2x^2 - x = 3$

$2x^2 - x - 3 = 0$

$(2x - 3)(\boxed{}) = 0$

$2x - 3 = 0$ *or* $x + 1 = 0$

$2x = 3$ *or* $x = -1$

$x = \boxed{}$ *or* $x = -1$

The solutions are $\boxed{}$

and $\boxed{}$.

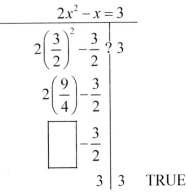

Check:

For $x = -1$:

$$\frac{2x^2 - x = 3}{2\left(\boxed{}\right)^2 - \left(\boxed{}\right) \overset{?}{} 3}$$

$$\begin{array}{c|c} 2 + 1 & \\ 3 & 3 \end{array} \quad \text{TRUE}$$

For $x = \dfrac{3}{2}$:

$$\frac{2x^2 - x = 3}{2\left(\dfrac{3}{2}\right)^2 - \dfrac{3}{2} \overset{?}{} 3}$$

$$2\left(\dfrac{9}{4}\right) - \dfrac{3}{2}$$

$$\boxed{} - \dfrac{3}{2}$$

$$\begin{array}{c|c} 3 & 3 \end{array} \quad \text{TRUE}$$

Example 2   Solve: $2x^2 - 10 = 0$.

$$2x^2 - 10 = 0$$
$$2x^2 = 10$$
$$x^2 = 5$$
$$x = \boxed{}, \text{ or } \sqrt{5} \text{ and } -\sqrt{5}$$

Check:

$$\begin{array}{c|c} \multicolumn{2}{c}{2x^2 - 10 = 0} \\ \hline 2\left(\boxed{}\right)^2 - 10 \ ? \ 0 & \\ 10 - 10 & \\ 0 & 0 \quad \text{TRUE} \end{array}$$

The solutions are $\sqrt{5}$ and $-\sqrt{5}$, or $\boxed{}$.

Example 3   Find the zeros of $f(x) = x^2 - 6x - 10$ by completing the square.

$$x^2 - 6x - 10 = 0$$
$$x^2 - 6x \qquad = 10$$
$$x^2 - 6x + \boxed{} = 10 + \boxed{}$$
$$\left(\boxed{}\right)^2 = 19$$
$$x - 3 = \boxed{}$$
$$x = 3 \pm \sqrt{19}$$

$$\frac{1}{2}\left(\boxed{}\right) = -3$$
$$(-3)^2 = \boxed{}$$

$$x = 3 + \sqrt{19} \quad or \quad x = 3 - \sqrt{19}$$
$$x \approx 7.359 \qquad or \quad x \approx -1.359$$

Zeros of $f(x) = x^2 - 6x - 10$ are

$$3 + \sqrt{19} \text{ and } \boxed{},$$

or approximately

$$\boxed{} \text{ and } -1.359.$$

Example 4   Solve: $2x^2 - 1 = 3x$.

$$2x^2 - 1 = 3x$$

$$2x^2 - \boxed{} - 1 = 0$$

$$2x^2 - 3x \quad\quad = 1$$

$$x^2 - \frac{3}{2}x \quad\quad = \boxed{}$$

$$x^2 - \frac{3}{2}x + \boxed{} = \frac{1}{2} + \boxed{} \quad\quad \text{Completing the square}$$

$$\left(\boxed{}\right)^2 = \frac{17}{16}$$

$$x - \frac{3}{4} = \pm\sqrt{\frac{17}{16}} = \pm\frac{\sqrt{17}}{\sqrt{16}} = \pm\frac{\sqrt{17}}{\boxed{}}$$

$$x = \frac{3}{4} \pm \frac{\sqrt{17}}{4}$$

$$x = \frac{3 \pm \sqrt{17}}{4}$$

The solutions are $\dfrac{3 + \sqrt{17}}{4}$ and $\dfrac{3 - \sqrt{17}}{4}$, or $\boxed{}$.

The Quadratic Formula

The solutions of $ax^2 + bx + c = 0$, $a \neq 0$, are given by

$$x = \frac{-b \pm \sqrt{b^2 - 4ac}}{2a}.$$

Example 5   Solve $3x^2 + 2x = 7$. Find exact solutions and approximate solutions rounded to three decimal places.

$$3x^2 + 2x = 7$$

$$3x^2 + 2x - 7 = 0$$

$$a = 3 \quad b = \boxed{} \quad c = \boxed{}$$

$$x = \frac{-b \pm \sqrt{b^2 - 4ac}}{2a}$$

$$x = \frac{-\boxed{} \pm \sqrt{2^2 - 4 \cdot 3 \cdot \left(\boxed{}\right)}}{2(3)}$$

$$= \frac{-2 \pm \sqrt{4 + 84}}{6} = \frac{-2 \pm \sqrt{\boxed{}}}{6}$$

$$= \frac{-2 \pm \sqrt{4 \cdot \boxed{}}}{6} = \frac{-2 \pm 2\sqrt{22}}{6}$$

$$= \frac{2\left(-1 \pm \sqrt{22}\right)}{2 \cdot 3} = \frac{\boxed{} \pm \sqrt{22}}{3}$$

Exact solutions: $\dfrac{-1 - \sqrt{22}}{3}$ and $\dfrac{-1 + \sqrt{22}}{3}$

Approximate solutions: -1.897 and $\boxed{}$

Example 6   Solve: $x^2 + 5x + 8 = 0.$

$a = \boxed{} \qquad b = 5 \qquad c = 8$

$$x = \frac{-b \pm \sqrt{b^2 - 4ac}}{2a}$$

$$x = \frac{-\boxed{} \pm \sqrt{5^2 - 4 \cdot \boxed{} \cdot \boxed{}}}{2 \cdot 1}$$

$$x = \frac{-5 \pm \sqrt{\boxed{}}}{2}$$

$$x = \frac{-5 \pm \sqrt{7}\,\boxed{}}{2}$$

$$x = -\frac{5}{2} + \frac{\sqrt{7}}{2}i \text{ or } x = -\frac{5}{2} - \frac{\sqrt{2}}{2}i$$

Discriminant

For $ax^2 + bx + c = 0$, where a, b, and c are real numbers, $a \neq 0$:

$\qquad b^2 - 4ac = 0 \rightarrow$ One real-number solution;

$\qquad b^2 - 4ac > 0 \rightarrow$ Two different real-number soluions;

$\qquad b^2 - 4ac < 0 \rightarrow$ Two different imaginary-number solutions,
$\qquad\qquad\qquad$ complex conjugates.

Example 7   Solve: $x^4 - 5x^2 + 4 = 0$.

Let $u = \boxed{}$.

$u^2 - 5\boxed{} + 4 = 0$

$(u-1)\left(\boxed{}\right) = 0$

$u - 1 = 0$ *or* $\boxed{} = 0$

$\quad u = 1$ *or* $\quad u = 4$

$x^2 = \boxed{}$ *or* $x^2 = 4$

$\quad x = \pm 1$ *or* $\quad x = \boxed{}$

The solutions are -1, 1, -2, and 2.

Example 8   Solve: $t^{2/3} - 2t^{1/3} - 3 = 0$.

Let $u = t^{1/3}$.

$\left(t^{1/3}\right)^{\boxed{}} - 2t^{1/3} - 3 = 0$

$\quad u^2 - 2\boxed{} - 3 = 0$

$\quad \left(\boxed{}\right)(u - 3) = 0$

$\quad u + 1 = 0$ *or* $u - 3 = 0$

$\quad\quad u = \boxed{}$ *or* $\quad u = 3$

$\boxed{} = -1$ *or* $\quad t^{1/3} = 3$

$\left(t^{1/3}\right)^3 = (-1)^3$ *or* $\left(t^{1/3}\right)^3 = 3^3$

$\boxed{} = -1$ $\quad\quad\quad t = 27$

The solutions are $\boxed{}$ and $\boxed{}$.

Example 9   *Museums in China.* The number of museums in China increased from approximately 2000 in the year 2000 to over 3500 by the end of 2012. In 2012, a record 451 new museums opened. For comparison, in the United States, only 20-40 new museums were opened per year from 2000 to 2008. The function

$$h(x) = 30.992x^2 + 4.108x + 2294.594$$

can be used to estimate the number of museums in China, *x* years after 2005.

a) Estimate the number of museums that will be in China in 2017 if the number of new museums that open per year continues at the same rate.

$$h(12) = 30.992\left(\boxed{}\right)^2 + 4.108\left(\boxed{}\right) + 2294.594$$

$$h(12) \approx 6807$$

In the year 2017, there are approximately $\boxed{}$ museums in China.

b) In what year was the number of museums in China 2600?

$$\boxed{} = 30.992x^2 + 4.108x + 2294.594$$

$$0 = 30.992x^2 + 4.108x - \boxed{}$$

$$a = 30.992 \quad b = 4.108 \quad c = -305.406$$

$$x = \frac{-b \pm \sqrt{b^2 - 4ac}}{2a}$$

$$x = \frac{-\boxed{} \pm \sqrt{(4.108)^2 - 4(30.992)(-305.406)}}{2(30.992)}$$

$$x = \frac{-4.108 \pm \sqrt{37,877.4467}}{61.984}$$

$$x = \boxed{} \quad or \quad x = \boxed{}$$

We are looking for a year after 2005, so we use the positive solution 3.074. Thus, there were about 2600 museums in China 3 years after 2005, or in $\boxed{}$.

Example 10   The numbers of both magazine launches and magazine closures have increased in recent years. The function $m(x) = 34x^2 - 59x + 81$ can be used to estimate the number of magazine closures after 2012. In what year was the number of magazine closures 99?

$$99 = 34x^2 - 59x + 81$$

$$0 = 34x^2 - 59x - \boxed{}$$

We use the quadratic formula with $a = 34$, $b = \boxed{}$, and $c = -18$.

$$x = \frac{-b \pm \sqrt{b^2 - 4ac}}{2a}$$

$$x = \frac{-(-59) \pm \sqrt{(-59)^2 - 4(34)(-18)}}{2(\boxed{})}$$

$$x = \frac{59 \pm \sqrt{5929}}{\boxed{}}$$

$$x = \frac{59 + 77}{68} \quad or \quad x = \frac{59 - 77}{68}$$

$$x = \frac{136}{68} \quad or \quad x = \frac{-18}{68}$$

$$x = \boxed{} \quad or \quad x \approx -0.3$$

We are looking for a year after 2012, so x must be positive. Thus there were 99 magazine closures 2 years after 2012, or in $\boxed{}$.

Example 11   *Train Speeds*. Two trains leave a station at the same time. One train travels due west, and the other travels due south. The train traveling west travels 20 km/h faster than the train traveling south. After 2 hr, the trains are 200 km apart. Find the speed of each train.

We use $d = rt$ to find the distance each train traveled in 2 hr.

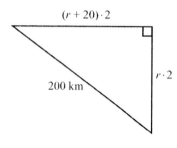

$(r + 20) \cdot 2$

$r \cdot 2$

200 km

We use the Pythagorean theorem.

$$\left[2(r+20)\right]^2 + \left(\boxed{}\right)^2 = 200^2$$

$$\frac{\boxed{}}{4} + \frac{4r^2}{4} = \frac{40,000}{4}$$

$$(r+20)^2 + r^2 = 10,000$$

$$r^2 + 40r + \boxed{} + r^2 = 10,000$$

$$\frac{2r^2}{2} + \frac{40r}{2} - \frac{9600}{2} = 0$$

$$r^2 + 20r - 4800 = 0$$

$$\left(\boxed{}\right)(r - 60) = 0$$

$$r + 80 = 0 \qquad or \quad r - 60 = 0$$

$$r = -80 \quad or \qquad r = \boxed{}$$

The negative value of r doesn't make sense in the problem. We check 60.

Check:

$$160^2 + 120^2 = 200^2$$

$$40,000 = 40,000$$

The answer checks.

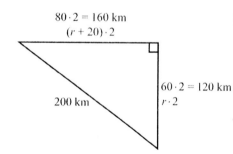

$80 \cdot 2 = 160$ km

$(r + 20) \cdot 2$

$60 \cdot 2 = 120$ km

$r \cdot 2$

200 km

The speed of the train heading south is $\boxed{}$ km/h, and the speed of the train heading west is $\boxed{}$ km/h.

Section 3.3 Analyzing Graphs of Quadratic Functions

Graphing Quadratic Functions

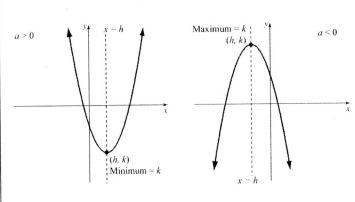

The graph of the function $f(x) = a(x-h)^2 + k$ is a parabola that

- opens up if $a > 0$ and down if $a < 0$;
- has (h, k) as the vertex;
- has $x = h$ as the axis of symmetry;
- has k as a minimum value (output) if $a > 0$;
- has k as a maximum value if $a < 0$.

Example 1   Find the vertex, the axis of symmetry, and the maximum or minimum value of $f(x) = x^2 + 10x + 23$. Then graph the function.

$$f(x) = a(x-h)^2 + k$$
$$f(x) = \left(x^2 + 10x + \boxed{}\right) - \boxed{} + 23$$
$$= (x+5)^2 - 2$$
$$= \left[x - \left(\boxed{}\right)\right]^2 + \left(\boxed{}\right)$$

Vertex: $(-5, -2)$

Axis of symmetry: $x = \boxed{}$

Minimum: -2

Some points on the graph:

$$(-5, -2)$$
$$(-4, -1)$$
$$(-2, 7)$$
$$(-7, 2)$$
$$(-8, 7)$$

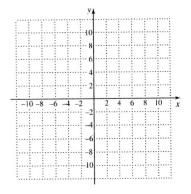

Example 2   Find the vertex, the axis of symmetry, and the maximum or minimum value of $g(x) = \dfrac{x^2}{2} - 4x + 8$. Then graph the function.

$$g(x) = a(x-h)^2 + k$$

$$g(x) = \dfrac{x^2}{2} - 4x + 8 = \dfrac{1}{2}\left(x^2 - 8x\right) + 8$$

$$= \dfrac{1}{2}\left(x^2 - 8x + \boxed{} - \boxed{}\right) + 8$$

$$= \dfrac{1}{2}\left(x^2 - 8x + 16\right) - \dfrac{1}{2}\cdot 16 + 8$$

$$= \dfrac{1}{2}\left(\boxed{}\right)^2 + \boxed{}, \text{ or } \dfrac{1}{2}(x-4)^2$$

Vertex: $(4, 0)$

Axis of symmetry: $x = \boxed{}$

Minimum: $\boxed{}$

Some points on the graph:

$(4, 0)$
$(0, 8)$
$(2, 2)$
$(6, 2)$
$(8, 8)$

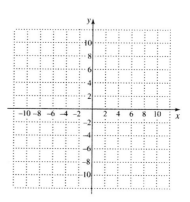

Example 3   Find the vertex, the axis of symmetry, and the maximum or minimum value of $f(x) = -2x^2 + 10x - \dfrac{23}{2}$. Then graph the function.

$$f(x) = -2x^2 + 10x - \dfrac{23}{2}$$

$$= -2\left(\boxed{}\right) - \dfrac{23}{2}$$

$$= -2\left(x^2 - 5x + \dfrac{25}{4} - \dfrac{25}{4}\right) - \dfrac{23}{2} \qquad \left(-\dfrac{5}{2}\right)^2 = \dfrac{25}{4}$$

$$= -2\left(x^2 - 5x + \dfrac{25}{4}\right) + \boxed{} - \dfrac{23}{2}$$

$$f(x) = -2\left(\boxed{}\right)^2 + \boxed{}$$

Vertex: $\left(\dfrac{5}{2}, 1\right)$

Axis of symmetry: $x = \boxed{}$

Maximum value is $\boxed{}$ when $x = \dfrac{5}{2}$.

Some points on the graph:

$\left(\dfrac{5}{2}, 1\right)$

$\left(2, \dfrac{1}{2}\right)$

$\left(3, \dfrac{1}{2}\right)$

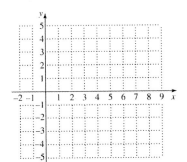

The Vertex of a Parabola

The vertex of the graph of $f(x) = ax^2 + bx + c$ is

$$\left(-\dfrac{b}{2a}, \; f\left(-\dfrac{b}{2a}\right)\right).$$

We calculate the x-coordinate. Then we substitute to find the y-coordinate.

Example 4 For the function $f(x) = -x^2 + 14x - 47$:

a) Find the vertex.

$\quad a = -1 \qquad b = 14 \qquad c = -47$

$\quad$ Vertex: $\left(\dfrac{-14}{2(-1)}, \; \boxed{}\right) = \left(7, \; \boxed{}\right)$

$\quad f(7) = -(7)^2 + 14 \cdot 7 - 47$

$\qquad\quad = -49 + 98 - 47 = \boxed{}$

$\quad$ The vertex is $(7, 2)$.

b) Determine whether there is a maximum or a minimum value and find that value.

$\quad$ Since the coefficient of x^2 is negative, the function has a $\underline{}$ value at

$\qquad\qquad\qquad\qquad\qquad\qquad\quad$ maximum / minimum

$\quad$ the vertex.

$\quad$ Maximum value: $\boxed{}$

c) Find the range.

Vertex: $(7, 2)$

Maximum value: 2

Range: $\left(-\infty, \boxed{}\right]$

d) On what intervals is the function increasing? decreasing?

Increasing as we approach the vertex from the left: $\left(-\infty, \boxed{}\right)$

Decreasing as we move away from the vertex on the right: $\left(\boxed{}, \infty\right)$

Example 5   *Maximizing Area.* A landscaper has enough stone to enclose a rectangular koi pond next to an existing garden wall of the Englemans' house with 24 ft of stone wall. If the garden wall forms one side of the rectangle, what is the maximum area that the landscaper can enclose? What dimensions of the koi pond will yield this area?

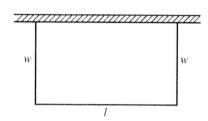

$2w + l = \boxed{}$

$l = 24 - 2w$

$A = l \cdot w$

$A = \left(\boxed{}\right) \cdot w$

$A = 24w - 2w^2$

$A = -2w^2 + 24w \qquad\qquad A(w) = aw^2 + bw + c$

Maximum occurs at the vertex:

$w = \dfrac{-b}{2a} = \dfrac{\boxed{}}{2(-2)} = \dfrac{-24}{-4} = \boxed{}$

$l = 24 - 2w = 24 - 2\left(\boxed{}\right) = 24 - 12 = \boxed{}$

$A = 12 \text{ ft} \cdot 6 \text{ ft} = 72 \text{ ft}^2$

The maximum possible area is $\boxed{}$ ft^2 when the koi pond is $\boxed{}$ ft wide and $\boxed{}$ ft long.

Example 6 *Height of a Rocket.* A model rocket is launched with an initial velocity of 100 ft/sec from the top of a hill that is 20 ft high. Its height, in feet, t seconds after it has been launched is given by the function $s(t) = -16t^2 + 100t + 20$. Determine the time at which the rocket reaches its maximum height and find the maximum height.

1. & 2.   Familiarize and Translate.

$$s(t) = -16t^2 + 100t + 20$$

3. Carry out.

$$a = -16 \qquad b = 100 \qquad c = 20$$

$$t = -\frac{b}{2a} = -\frac{\boxed{}}{2(-16)} = -\frac{100}{-32}$$

$$= \boxed{}$$

$$s(3.125) = -16\left(\boxed{}\right)^2 + 100\left(\boxed{}\right) + 20$$

$$= \boxed{}$$

4. Check. We can check the answer by using the MAX feature on a graphing calculator. We can also complete the square on the original function. If we did that, we would get

$$s(t) = -16(t - 3.125)^2 + 176.25.$$

This confirms that the vertex is $(3.125, 176.25)$.

5. State. The rocket reaches a maximum height of $\boxed{}$ ft $\boxed{}$ sec after it has been launched.

Example 7   Jared drops a screwdriver from the top of an elevator shaft. Exactly 5 sec later, he hears the sound of the screwdriver hitting the bottom of the shaft. The speed of sound is 1100 ft/sec. How tall is the elevator shaft?

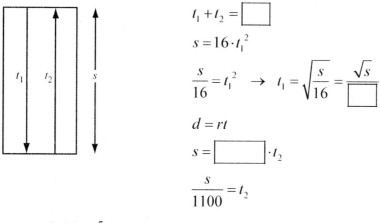

$$t_1 + t_2 = \boxed{}$$

$$s = 16 \cdot t_1^2$$

$$\frac{s}{16} = t_1^2 \quad \rightarrow \quad t_1 = \sqrt{\frac{s}{16}} = \frac{\sqrt{s}}{\boxed{}}$$

$$d = rt$$

$$s = \boxed{} \cdot t_2$$

$$\frac{s}{1100} = t_2$$

$$t_1 + t_2 = 5$$

$$\frac{\sqrt{s}}{4} + \boxed{} = 5$$

$$275\sqrt{s} + s = \boxed{}$$

$$s + 275\sqrt{s} - 5500 = 0$$

Let $u = \sqrt{s}$.

$$u^2 + 275u - 5500 = 0$$

$$u = \frac{-\boxed{} \pm \sqrt{275^2 - 4(1)(-5500)}}{2 \cdot 1}$$

$$u = \frac{-275 \pm \sqrt{97625}}{2}$$

$$u = \frac{-275 + \sqrt{97625}}{2} \qquad \text{We want only the positive solution.}$$

$$u \approx \boxed{}$$

$$\sqrt{s} \approx 18.725$$

$$s \approx 350.6 \text{ ft}$$

The height of the elevator shaft is about $\boxed{}$ ft.

Section 3.4 Solving Rational Equations and Radical Equations

> Equations containing rational expressions are called **rational equations**.

Example 1   Solve: $\dfrac{x-8}{3}+\dfrac{x-3}{2}=0$.

$LCD = 3\cdot 2 = \boxed{}$

$\boxed{}\cdot\left(\dfrac{x-8}{3}+\dfrac{x-3}{2}\right)=\boxed{}\cdot 0$

$2(x-8)+\boxed{}\cdot(x-3)=0$

$2x-\boxed{}+3x-9=0$

$5x-\boxed{}=0$

$5x=25$

$x=\boxed{}$

Example 2   Solve: $\dfrac{x^2}{x-3}=\dfrac{9}{x-3}$.

$LCD = x-3$

$\left(\boxed{}\right)\dfrac{x^2}{x-3}=(x-3)\dfrac{9}{x-3}$

$x^2=\boxed{}$

$x=\pm 3$

Check:

For -3:

$\dfrac{x^2}{x-3}=\dfrac{9}{x-3}$

$\dfrac{(-3)^2}{-3-3}\overset{?}{}\dfrac{9}{-3-3}$

$\dfrac{9}{-6}\;\bigg|\;\dfrac{9}{-6}$ TRUE

For 3:

$\dfrac{x^2}{x-3}=\dfrac{9}{x-3}$

$\dfrac{3^2}{3-3}\overset{?}{}\dfrac{9}{3-3}$

$\dfrac{9}{0}\;\bigg|\;\dfrac{9}{0}$ Not defined

The number $\boxed{}$ is a solution. Since division by 0 is not defined, $\boxed{}$ is not a solution.

Example 3   Solve: $\dfrac{2}{3x+6} + \dfrac{1}{x^2-4} = \dfrac{4}{x-2}$.

$$\frac{2}{3(x+2)} + \frac{1}{\boxed{}(x-2)} = \frac{4}{x-2} \qquad \text{LCD} = 3(x+2)(\boxed{})$$

$$3(x+2)(x-2)\left(\frac{2}{3(x+2)} + \frac{1}{(x+2)(x-2)}\right) = 3(x+2)(x-2)\cdot\frac{4}{x-2}$$

$$2\left(\boxed{}\right) + 3 = \boxed{}\cdot(x+2)$$

$$2x - \boxed{} + 3 = 12x + 24$$

$$2x - 1 = 12x + 24$$

$$\boxed{} = 25$$

$$x = \frac{25}{-10}$$

$$x = \boxed{}$$

A **radical equation** is an equation in which variables appear in one or more radicands.

The Principle of Powers
For any positive integer n: If $a = b$ is true, then $a^n = b^n$ is true.

Example 4   Solve: $\sqrt{3x+1} = 4$.

$$\left(\sqrt{3x+1}\right)^2 = 4^{\boxed{}}$$

$$\boxed{} = 16$$

$$3x = 15$$

$$x = \boxed{}$$

Check:

$$\begin{array}{c}\sqrt{3x+1} = 4 \\ \hline \sqrt{3\cdot 5+1}\;?\;4 \\ \sqrt{16}\;\Big|\; \\ 4\;\Big|\;4 \qquad \text{TRUE}\end{array}$$

The solution is $\boxed{}$.

Example 5   Solve: $5 + \sqrt{x+7} = x$.

$$\sqrt{x+7} = x - 5$$

$$\left(\sqrt{x+7}\right)^{\square} = (x-5)^2$$

$$x + 7 = x^2 - \boxed{} + 25$$

$$0 = x^2 - 11x + \boxed{}$$

$$0 = \left(\boxed{}\right)(x-2)$$

$$x - 9 = 0 \qquad or \quad x - 2 = 0$$

$$x = \boxed{} \quad or \qquad x = \boxed{}$$

Check:

For 9:

$$5 + \sqrt{x+7} = x$$

$$\overline{5 + \sqrt{9+7} \; ? \; 9}$$

$$5 + \sqrt{16}$$

$$5 + 4$$

$$9 \; \big| \; 9 \qquad \text{TRUE}$$

For 2:

$$5 + \sqrt{x+7} = x$$

$$\overline{5 + \sqrt{2+7} \; ? \; 2}$$

$$5 + \sqrt{9}$$

$$5 + 3$$

$$8 \; \big| \; 2 \qquad \text{FALSE}$$

The solution is $\boxed{}$.

Example 6   Solve: $\sqrt{x-3} + \sqrt{x+5} = 4$.

$$\left(\sqrt{x-3}\right)^2 = \left(4 - \sqrt{x+5}\right)^{\square}$$

$$x - 3 = 16 - \boxed{} \cdot \sqrt{x+5} + (x+5)$$

$$x - 3 = \boxed{} - 8\sqrt{x+5} + x$$

$$-24 = -8\sqrt{x+5}$$

$$3 = \sqrt{x+5}$$

$$3^{\square} = \left(\sqrt{x+5}\right)^{\square}$$

$$9 = x + 5$$

$$x = \boxed{}$$

The solution is 4.

Section 3.5   Solving Equations and Inequalities with Absolute Value

> For $a > 0$ and an algebraic expression X:
> $|X| = a$ is equivalent to $X = -a$ or $X = a$.

Example 1   Solve: $|x| = 5$.

$x = \boxed{}$ *or* $x = \boxed{}$

The solutions are -5 and 5.

Example 2   Solve: $|x - 3| - 1 = 4$.

$|x - 3| = 5$

$x - 3 = \boxed{}$ *or* $x - 3 = \boxed{}$

$\quad x = -2 \quad$ *or* $\quad x = 8$

Check:

For -2:

$$|x - 3| - 1 = 4$$

$|-2 - 3| - 1 \overset{?}{} 4$

$|-5| - 1$

$5 - 1$

$\qquad 4 \mid 4 \qquad$ TRUE

For 8:

$$|x - 3| - 1 = 4$$

$|8 - 3| - 1 \overset{?}{} 4$

$|5| - 1$

$5 - 1$

$\qquad 4 \mid 4 \qquad$ TRUE

The solutions are -2 and $\boxed{}$.

For $a > 0$ and an algebraic expression X:

$|X| < a$ is equivalent to $-a < X < a$.

$|X| > a$ is equivalent to $X < -a$ or $X > a$.

Similar statements hold for $|X| \le a$ and $|X| \ge a$.

Example 3   Solve and graph the solution set: $|3x + 2| < 5$.

$\boxed{} < 3x + 2 < \boxed{}$

$-7 < 3x < 3$

$-\dfrac{7}{3} < x < \boxed{}$

$\left\{ x \middle| \boxed{} < x < 1 \right\}$, or $\left(-\dfrac{7}{3}, \boxed{} \right)$

Example 4   Solve and graph the solution set: $|5 - 2x| \ge 1$.

$5 - 2x \le \boxed{}$ or $5 - 2x \ge 1$

$-2x \le -6$ or $-2x \ge -4$

$x \boxed{} 3$ or $x \boxed{} 2$

$\{x | x \le 2 \ or \ x \ge 3\}$, $\left(\boxed{}, 2 \right] \cup \left[\boxed{}, \infty \right)$

Section 4.1   Polynomial Functions and Modeling

Polynomial Function

A **polynomial function** P is given by

$$P(x) = a_n x^n + a_{n-1} x^{n-1} + a_{n-2} x^{n-2} + \ldots + a_1 x + a_0,$$

where the coefficients a_n, a_{n-1}, ..., a_1, a_0 are real numbers and the exponents are whole numbers.

The Leading-Term Test

If $a_n x^n$ is the leading term of a polynomial function, then the behavior of the graph as $x \to \infty$ or as $x \to -\infty$ can be described in one of the four following ways:

The ⋙ portion of the graph is not determined by this test.

Example 1   Use the leading-term test to match the functions below with one of the graphs on the next page.

a) $f(x) = 3x^4 - 2x^3 + 3$

Exponent: <u>even / odd</u> (circle one)
Coefficient: <u>positive / negative</u> (circle one)
Graph: <u>A, B, C, or D</u> (circle one)

b) $f(x) = -5x^3 - x^2 + 4x + 2$

Exponent: <u>even / odd</u> (circle one)
Coeffieicnt: <u>positive / negative</u> (circle one)
Graph: <u>A, B, C, D</u> (circle one)

c) $f(x) = x^5 + \dfrac{1}{4}x + 1$

Exponent: <u>even / odd</u> (circle one)
Coefficient: <u>positive / negative</u> (circle one)
Graph: <u>A, B, C, D</u> (circle one)

d) $f(x) = -x^6 + x^5 - 4x^3$

Exponent: <u>even / odd</u> (circle one)
Coefficient: <u>positive / negative</u> (circle one)
Graph: <u>A, B, C, D</u> (circle one)

A. 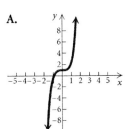   B. 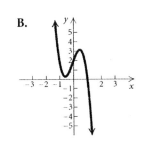   C. 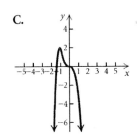   D.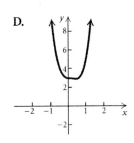

Example 2   Consider $P(x) = x^3 + x^2 - 17x + 15$. Determine whether each of the numbers 2 and -5 is a zero of $P(x)$.

If $P(c) = 0$, then c is a zero of the polynomial.

Let us first evaluate $P(2)$.

$P(2) = 2^3 + 2^2 - 17 \cdot 2 + 15$

$P(2) = 8 + \boxed{} - 34 + 15$

$P(2) = \boxed{}$

Since $P(2) \neq 0$, 2 <u>is / is not</u> (circle one) a zero of this function.

Let us now evaluate $P(-5)$.

$P(-5) = (-5)^3 + (-5)^2 - 17 \cdot (-5) + 15$

$P(-5) = -125 + 25 + 85 + 15$

$P(-5) = 0$

Since $P(-5) = 0$, -5 <u>is / is not</u> (circle one) a zero of this function.

Example 3   Find the zeros of

$$f(x) = 5(x-2)(x-2)(x-2)(x+1)$$
$$= 5(x-2)^3(x+1).$$

According to the principle of zero products, either

$x - 2 = 0$ *or* $x + 1 = 0$

$x = 2$ *or* $x = -1$ So the zeros of $f(x)$ are $\boxed{}$ and $\boxed{}$.

Example 4   Find the zeros of

$$g(x) = -(x-1)(x-1)(x+2)(x+2)$$
$$= -(x-1)^2(x+2)^2.$$

To find the zeros, we set $g(x) = 0$, and use the principle of zero products.

$0 = -(x-1)^2(x+2)^2$

$x - 1 = 0$ *or* $x + 2 = 0$

$x = 1$ *or* $x = -2$ The zeros are $\boxed{}$ and $\boxed{}$.

Example 5   Find the zeros of $f(x) = x^3 - 2x^2 - 9x + 18$.

$$f(x) = x^3 - 2x^2 - 9x + 18$$
$$= x^2\left(x - \boxed{}\right) - 9\left(\boxed{} - 2\right)$$
$$= (x - 2)(x^2 - 9)$$
$$= (x - 2)(x + 3)\left(x - \boxed{}\right)$$

To find the zeros, we solve the equation $f(x) = 0$ using the principle of zero products.

$$0 = (x - 2)(x + 3)(x - 3)$$
$$x - 2 = 0 \quad or \quad x + 3 = 0 \quad or \quad x - 3 = 0$$
$$x = 2 \quad or \quad\quad x = -3 \quad or \quad\quad x = 3$$

The solutions of $f(x) = 0$ are 2, $\boxed{}$ and 3.

The zeros of $f(x)$ are $\boxed{}$, -3 and $\boxed{}$.

Example 6   Find the zeros of $f(x) = x^4 + 4x^2 - 45$.

To find the zeros, we solve the equation $f(x) = 0$.

$$0 = x^4 + 4x^2 45$$
$$0 = \left(x^2 - \boxed{}\right)(x^2 + 9)$$

Using the principle of zero products, this becomes

$$x^2 - 5 = 0 \quad or \quad x^2 + 9 = 0$$
$$x^2 = 5 \quad or \quad\quad x^2 = -9$$
$$x = \pm\sqrt{5} \quad or \quad\quad x = \pm\sqrt{-9}$$
$$x = \pm 3i$$

The solutions of $f(x) = 0$ are $-\sqrt{5}$, $\sqrt{5}$, $\boxed{}$, and $3i$.

The zeros of $f(x)$ are $-\sqrt{5}$, $\boxed{}$, $-3i$, and $\boxed{}$.

Example 7   Find the real zeros of the function given by

$$f(x) = 0.1x^3 - 0.6x^2 - 0.1x + 2.$$

Approximate the zeros to three decimal places.

We use a graphing calculator to graph the function. We see that the graph intersects the x-axis at three points. There are no more than ☐ zeros because the degree of the function is ☐.

Use the ZERO feature to find the zeros. They are approximately -1.680, 2.154, and ☐.

Example 8   The polynomial function

$$M(t) = 0.5t^4 + 3.45t^3 - 96.65t^2 + 347.7t$$

can be used to estimate the number of milligrams of the pain relief medication ibuprofen in the bloodstream t hours after 400 mg of the medication has been taken.

a) Find the number of milligrams in the bloodstream at $t = 0$, 0.5, 1, 1.5, and so on, up to 6 hr. Round the function values to the nearest tenth.

Using a calculator, we compute the function values:

$M(0) = 0$, $M(3.5) = 255.9$,

$M(0.5) = 150.2$, $M(4) = \boxed{}$,

$M(1) = 255$, $M(4.5) = 126.9$,

$M(1.5) = 318.3$, $M(5) = 66$,

$M(2) = \boxed{}$, $M(5.5) = 20.2$,

$M(2.5) = 338.6$, $M(6) = 0$.

$M(3) = 306.9$,

b) Find the domain, the relative maximum and where it occurs, and the range.

Domain: We need to think about this application in finding the domain. We have $M(0) = 0$; then $M(t)$ is positive until it returns to 0 at $t = 6$. Thus, the domain is $\left[0, \boxed{}\right]$.

Relative maximum: Using the MAXIMUM feature on a graphing calculator, we find that the maximum is about $\boxed{}$ mg. It occurs approximately 2.15 hr, or 2 hr 9 min, after the dose was taken.

Range: $\left[0, \boxed{}\right]$

Example 9   *Army Personnel on Active Duty.* U.S. Army personnel on active duty make up approximately 37% of the total number in the U.S. military. Midyear numbers, in thousands, of U.S. Army personnel for selected years are listed in the following table.

Year, x	Number of Active-Duty Army Personnel (in thousands)
2005, 0	492.7
2007, 2	522.0
2008, 3	539.2
2010, 5	566.0
2012, 7	550.1
2014, 9	512.1

a) Model the data with a quadratic function, a cubic function, and a quartic function. Let the first coordinate of each data point be the number of years after 2005. Then using R^2, the coefficient of determination, decide which function is the best fit.

Enter the data in lists and use the REGRESSION feature with DIAGNOSTIC turned on.

Quadratic function: $f(x) = -2.695463547x^2 + 27.48096918x + 486.7387796$;
$R^2 = 0.9281065984$.

Cubic function:
$f(x) = -0.2824686269x^3 + 1.063162809x^2 + 15.42806538x + 491.7267161$;
$R^2 = 0.9810056029$

Quartic function:
$f(x) = 0.0928485698x^4 - 1.9470803551x^3 + 10.1724474x^2 + 0.6034480945x + 492.849438$;
$R^2 = 0.9979418568$

The R^2-value for the _____ function is closest to 1, so this function
                          quadratic / cubic / quartic

is the best fit.

b) Graph the function with the scatterplot of the data.

Paste the quartic regression function into the calculator as Y1. Turn on Plot1, then select a viewing window. One good choice is $[0, 11, 480, 580]$, $\text{Yscl} = 10$, and press GRAPH. (Sketch the graph in the space below.)

c) Use the answer to part (a) to estimate the number of active-duty Army personnel in 2006, in 2009, in 2013, and in 2020.

We use a table set in ASK mode to evaluate the quartic function for $x = 1$, 4, 8, and 15.

$f(1) \approx 501.8$ thousand, or 501,800

$f(4) \approx \boxed{}$ thousand, or 557,200

$f(8) \approx 532.1$ thousand, or $\boxed{}$

$f(15) \approx 919.8$ thousand, or 919,800

The estimates for 2006, 2009, and 2013 appear to be fairly accurate. The estimate for 2020 is not realistic, since it is not reasonable to expect the number of active-duty Army personnel to increase by 407,700 from 2014 to 2020.

Section 4.2   Graphing Polynomial Functions

> **To Graph a Polynomial Function:**
> 1. Use the leading-term test to determine the end behavior.
> 2. Find the zeros of the function by solving $f(x) = 0$. Any real zeros are the first coordinates of the x-intercepts.
> 3. Use the x-intercepts (zeros) to divide the x-axis into intervals and choose a test point in each interval to determine the sign of all function values in that interval.
> 4. Find $f(0)$. This gives the y-intercept of the function.
> 5. If necessary, find additional function values to determine the general shape of the graph and then draw the graph.
> 6. As a partial check, use the facts that the graph has at most n x-intercepts and at most $n - 1$ turning points. Multiplicity of zeros can also be considered in order to check where the graph crosses or is tangent to the x-axis. We can also check the graph with a graphing calculator.

Example 1   Graph the polynomial function $h(x) = -2x^4 + 3x^3$.

1. Use the leading-term test.
   Degree: 4, even; coefficient: $-2 < 0$
   As $x \to \infty$ and as $x \to -\infty$, $h(x) \to \boxed{}$.

2. Find the zeros. Solve $h(x) = 0$.

$$-2x^4 + 3x^3 = 0$$

$$-x^3(2x - 3) = 0$$

$$\boxed{} = 0 \quad or \quad 2x - 3 = 0$$

$$x = 0 \quad or \quad x = \boxed{}$$

3.

   Choose a test value from each interval and find $h(x)$.

Interval	$(-\infty, 0)$	$\left(0, \dfrac{3}{2}\right)$	$\left(\dfrac{3}{2}, \infty\right)$
Test value	-1	1	2
Function value, $h(x)$	-5	1	-8
Sign of $h(x)$	$-$	$+$	$-$
Location of points on graph	Above/Below x-axis (circle one)	Above/Below x-axis (circle one)	Above/Below x-axis (circle one)

Three points on the graph are $\left(-1,\boxed{}\right)$, $(1,1)$ and $\left(\boxed{},-8\right)$.

4. To determine the y-intercept, find $h(0)$.

$$h(x) = -2x^4 + 3x^3$$
$$h(0) = -2 \cdot 0^4 + 3 \cdot 0^3 = \boxed{}$$

y-intercept: $(0,0)$

5. Find additional function values and draw the graph.

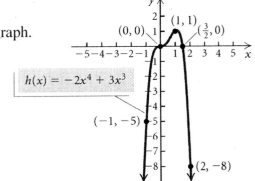

x	$h(x)$
-1.5	-20.25
-0.5	-0.5
0.5	$\boxed{}$
2.5	-31.25

Example 2 Graph the polynomial function $f(x) = 2x^3 + x^2 - 8x - 4$.

1. Use the leading-term test.

   Degree: 3, odd; coefficient: $2 > 0$

   As $x \to \infty$, $f(x) \to \infty$; as $x \to -\infty$, $f(x) \to -\infty$.

2. Find the zeros. Solve $f(x) = 0$.

$$2x^3 + x^2 - 8x - 4 = 0$$
$$x^2(2x+1) - 4(2x+1) = 0$$
$$(2x+1)\left(x^2 - \boxed{}\right) = 0$$
$$(2x+1)(x+2)(x-2) = 0$$

$\boxed{} = 0$ *or* $x + 2 = 0$ *or* $x - 2 = 0$

$x = -\dfrac{1}{2}$ *or* $x = \boxed{}$ *or* $x = 2$

3.

$$-5 \; -4 \; -3 \; -2 \; -1 \; \; 0 \; \; 1 \; \; 2 \; \; 3 \; \; 4 \; \; 5 \quad x$$

with marks at -2, $-\frac{1}{2}$, 2

Choose a test value from each interval and find $f(x)$.

Interval	$(-\infty, -2)$	$\left(\boxed{}, -\frac{1}{2}\right)$	$\left(-\frac{1}{2}, 2\right)$	$\left(\boxed{}, \infty\right)$
Test value	-3	-1	1	3
Function value, $f(x)$	-25	3	-9	35
Sign of $f(x)$	$-$	$+$	$-$	$+$
Location of points on graph	Above/Below x-axis (circle one)	Above/Below x-axis (circle one)	Above/Below x-axis (circle one)	Above/Below x-axis (circle one)

Four points on the graph are $\left(-3, -25\right)$, $\left(-1, \boxed{}\right)$, $\left(1, -9\right)$ and $\left(\boxed{}, 35\right)$.

4. To determine the y-intercept, find $f(0)$.

$$f(x) = 2x^3 + x^2 - 8x - 4$$

$$f(0) = 2 \cdot 0^3 + 0^2 - 8 \cdot 0 - 4 = \boxed{}$$

y-intercept: $\left(0, \boxed{}\right)$

5. Find additional function values and draw the graph.

x	$f(x)$
-2.5	-9
-1.5	3.5
0.5	-7.5
1.5	-7

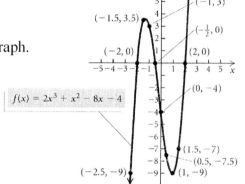

Example 3   Graph the polynomial function

$$g(x) = x^4 - 7x^3 + 12x^2 + 4x - 16$$

$$= (x+1)(x-2)^2(x-4).$$

1. Use the leading-term test.

Degree: 4, even; coefficient: $1 > 0$

As $x \to \infty$ and as $x \to -\infty$, $g(x) \to \infty$.

2. Find the zeros. Solve $g(x) = 0$.

$$(x+1)(x-2)^2(x-4) = 0$$

$$x + 1 = 0 \quad or \quad (x-2)^2 = 0 \quad or \quad x - 4 = 0$$

$$x = -1 \quad or \quad \qquad x = 2 \quad or \quad \qquad x = 4$$

3.

Choose a test value from each interval and find $g(x)$.

Interval	$(-\infty, -1)$	$(\boxed{}, 2)$	$(2, 4)$	$(\boxed{}, \infty)$
Test value	-1.25	1	3	4.25
Function value, $g(x)$	≈ 13.9	-6	-4	≈ 6.6
Sign of $g(x)$	$+$	$-$	$-$	$+$
Location of points on graph	Above/Below x-axis (circle one)	Above/Below x-axis (circle one)	Above/Below x-axis (circle one)	Above/Below x-axis (circle one)

4. To determine the y-intercept, find $g(0)$.

$$g(x) = x^4 - 7x^3 + 12x^2 + 4x - 16$$
$$g(0) = 0^4 - 7 \cdot 0^3 + 12 \cdot 0^2 + 4 \cdot 0 - 16 = \boxed{}$$

y-intercept: $\left(\boxed{}, -16\right)$

5. Find additional function values and draw the graph.

x	$g(x)$
-0.5	-14.1
0.5	-11.8
1.5	-1.6
2.5	-1.3
3.5	-5.1

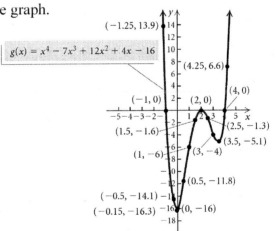

The Intermediate Value Theorem

For any polynomial function $P(x)$ with real coefficients, suppose that for $a \neq b$, $P(a)$ and $P(b)$ are of opposite signs. Then the function has a real zero between a and b.

The intermediate value theorem *cannot* be used to determine whether there is a real zero between a and b when $P(a)$ and $P(b)$ have the *same* sign.

Example 4

a) Using the intermediate value theorem, determine, if possible, whether
$f(x) = x^3 + x^2 - 6x$ has a real zero between -4 and -2.

$$f(-4) = (-4)^3 + (-4)^2 - 6(-4) = \boxed{}$$

$$f(-2) = (-2)^3 + (-2)^2 - 6(-2) = \boxed{}$$

Because $f(-4)$ and $f(-2)$ have opposite signs, $f(x)$ has / does not have (circle one) a
zero between -4 and -2.

b) Using the intermediate value theorem, determine, if possible, whether
$f(x) = x^3 + x^2 - 6x$ has a real zero between -1 and 3.

$$f(-1) = (-1)^3 + (-1)^2 - 6(-1) = \boxed{}$$

$$f(3) = (3)^3 + (3)^2 - 6(3) = \boxed{}$$

Because $f(-1)$ and $f(3)$ have the same sign, it is / is not (circle one) possible to
determine with the intermediate value theorem if there is a zero between -1 and 3.

c) Using the intermediate value theorem, determine, if possible, whether $g(x) = \frac{1}{3}x^4 - x^3$

has a real zero between $-\frac{1}{2}$ and $\frac{1}{2}$.

$$g\left(-\frac{1}{2}\right) = \frac{1}{3}\left(-\frac{1}{2}\right)^4 - \left(-\frac{1}{2}\right)^3 = \boxed{}$$

$$g\left(\frac{1}{2}\right) = \frac{1}{3}\left(\frac{1}{2}\right)^4 - \left(\frac{1}{2}\right)^3 = \boxed{}$$

Because $g\left(-\frac{1}{2}\right)$ and $g\left(\frac{1}{2}\right)$ have opposite signs, $g(x)$ has / does not have (circle one) a

zero between $-\frac{1}{2}$ and $\frac{1}{2}$.

d) Using the intermediate value theorem, determine, if possible, whether $g(x) = \frac{1}{3}x^4 - x^3$

has a real zero between 1 and 2.

$$g(1) = \frac{1}{3}(1)^4 - (1)^3 = \boxed{}$$

$$g(2) = \frac{1}{3}(2)^4 - (2)^3 = \boxed{}$$

Because $g(1)$ and $g(2)$ have the same sign, it is / is not (circle one) possible to
determine with the intermediate value theorem if there is a zero between 1 and 2.

Section 4.3 Polynomial Division; The Remainder Theorem and the Factor Theorem

Example 1   Divide to determine whether $x+1$ and $x-3$ are factors of $x^3 + 2x^2 - 5x - 6$.

First Divisor: $x+1$

$$
\begin{array}{r}
\boxed{} \\
x+1 \,) \overline{x^3 \;+\; 2x^2 \;-\; 5x \;-\; 6} \\
\underline{x^3 \;+\; \boxed{}} \\
x^2 \;-\; 5x \\
\underline{x^2 \;+\; x} \\
\boxed{} \;-\; 6 \\
\underline{-6x \;-\; 6} \\
\boxed{} \quad \leftarrow \text{Remainder}
\end{array}
$$

Since the remainder is 0, we know that $x+1$ <u>is / is not</u> (circle one) a factor of $x^3 + 2x^2 - 5x - 6$.

Second Divisor: $x-3$

$$
\begin{array}{r}
\boxed{} \\
x-3 \,) \overline{x^3 \;+\; 2x^2 \;-\; 5x \;-\; 6} \\
\underline{x^3 \;-\; 3x^2} \\
5x^2 \;-\; 5x \\
\underline{\boxed{} \;-\; 15x} \\
10x \;-\; \boxed{} \\
\underline{10x \;-\; 30} \\
\boxed{} \quad \leftarrow \text{Remainder}
\end{array}
$$

Since the remainder is 24, we know that $x-3$ <u>is / is not</u> (circle one) a factor of $x^3 + 2x^2 - 5x - 6$.

Example 2 Use synthetic division to find the quotient and the remainder:
$$(2x^3 + 7x^2 - 5) \div (x + 3).$$

First, rewrite the dividend including the missing x-term, and write the divisor in the form $(x - c)$:
$$(2x^3 + 7x^2 + 0x - 5) \div (x - (-3))$$

Now perform synthetic division:

$$\begin{array}{r|rrrr} -3 & 2 & 7 & 0 & -5 \\ & & -6 & \boxed{} & 9 \\ \hline & 2 & \boxed{} & -3 & \boxed{} \end{array}$$

The quotient is $2x^2 + x - 3$. The remainder is $\boxed{}$.

The Remainder Theorem

If a number c is substituted for x in the polynomial $f(x)$, then the result $f(c)$ is the remainder that would be obtained by dividing $f(x)$ by $x - c$.

That is, if $f(x) = (x - c) \cdot Q(x) + R$, then $f(c) = R$.

Example 3 Given that $f(x) = 2x^5 - 3x^4 + x^3 - 2x^2 + x - 8$, find $f(10)$.

$f(10)$ is the remainder when $f(x)$ is divided by $x - 10$.

We can use synthetic division to find $f(10)$.

$$\begin{array}{r|rrrrrr} 10 & 2 & -3 & 1 & -2 & 1 & -8 \\ & & 20 & \boxed{} & 1710 & 17{,}080 & 170{,}810 \\ \hline & 2 & 17 & 171 & \boxed{} & 17{,}081 & \boxed{} \end{array}$$

Thus, $f(10) = \boxed{}$.

Example 4   Determine whether 5 is a zero of $g(x)$, where $g(x) = x^4 - 26x^2 + 25$.

If 5 is a zero of $g(x)$, then $g(5) = 0$. If we divide $g(x)$ by $x - 5$ and get a remainder of 0, then $g(5) = 0$ and 5 is a zero of the function.

We use synthetic division to find $g(5)$, remembering to write zeros for the missing terms in $g(x)$.

$$
\begin{array}{r|rrrrr}
5 & 1 & 0 & -26 & 0 & 25 \\
  &   & 5 & \boxed{} & -5 & -25 \\
\hline
  & \boxed{} & 5 & -1 & -5 & \boxed{}
\end{array}
$$

$g(5) = \boxed{}$, so 5 is a zero of $g(x)$.

Example 5   Determine whether i is a zero of $f(x)$, where $f(x) = x^3 - 3x^2 + x - 3$.

If i is a zero of $f(x)$, then $f(i) = 0$. If we divide $f(x)$ by $x - i$ and get a remainder of 0, then $f(i) = 0$ and i is a zero of the function.

We use synthetic division to find $f(i)$.

$$
\begin{array}{r|rrrr}
i & 1 & -3 & 1 & -3 \\
  &   & \boxed{} & -1-3i & 3 \\
\hline
  & 1 & -3+i & \boxed{} & \boxed{}
\end{array}
$$

$f(i) = \boxed{}$, so i is a zero of $f(x)$.

The Factor Theorem

For a polynomial $f(x)$, if $f(c) = 0$, then $x - c$ is a factor of $f(x)$.

Example 6   Let $f(x) = x^3 - 3x^2 - 6x + 8$. Factor $f(x)$ and solve the equation $f(x) = 0$.

According to the factor theorem, if $f(c) = 0$, then $x - c$ is a factor of $f(x)$. At this point, we make some arbitrary choices for c and determine if they are zeros.

First try $c = -1$. This means we are dividing by $x + 1$.

$$
\begin{array}{r|rrrr}
-1 & 1 & -3 & -6 & 8 \\
   &   & -1 & \square & \square \\
\hline
   & 1 & -4 & -2 & \boxed{} \\
\end{array}
$$

$10 \neq 0$, so -1 is not a zero of $f(x)$. That means that $x + 1$ is not a factor of $f(x)$.

Now try $c = 1$. This tests whether $x - 1$ is a factor of $f(x)$.

$$
\begin{array}{r|rrrr}
1 & 1 & -3 & -6 & 8 \\
  &   & 1 & -2 & -8 \\
\hline
  & 1 & \square & -8 & \boxed{} \\
\end{array}
$$

This tells us that one of the factors of $f(x)$ is $x - 1$. Another factor is the quotient that we found when dividing. We have

$$
\begin{aligned}
f(x) &= (x-1)(x^2 - 2x - 8) \\
     &= (x-1)(x-4)(x+2).
\end{aligned}
$$

Finally, we solve the equation $f(x) = 0$.

$$(x-1)(x-4)(x+2) = 0$$

$$x - 1 = 0 \quad or \quad x - 4 = 0 \quad or \quad x + 2 = 0$$

$$x = \boxed{} \quad or \quad x = \boxed{} \quad or \quad x = \boxed{}$$

The solutions are $\boxed{}$, $\boxed{}$, and $\boxed{}$.

Section 4.4   Theorems about Zeros of Polynomial Functions

In this section, we will need to use the following relationship between the factors of $f(x)$ and the zeros of $f(x)$.

The Factor Theorem

For a polynomial $f(x)$, if $f(c) = 0$, then $x - c$ is a factor of $f(x)$.

The Fundamental Theorem of Algebra

Every polynomial function of degree n, with $n \geq 1$, has at least one zero in the set of complex numbers.

We will also need to know one of the results of the fundamental theorem of algebra:

Every polynomial function f of degree n, with $n \geq 1$, can be factored into n linear factors (not necessarily unique); that is,

$$f(x) = a_n(x - c_1)(x - c_2) \ldots (x - c_n).$$

Example 1   Find a polynomial function of degree 3, having the zeros 1, $3i$, and $-3i$.

If these are the zeros, then the factors must be

$(x - 1)$, $(x - 3i)$, $(x - (-3i))$, or

$(x - 1)$, $(x - 3i)$, $(x + 3i)$.   Writing $(x - (-3i))$ as $(x + 3i)$

So the function must have the form

$$f(x) = a_n(x - 1)(x - 3i)(x + 3i).$$

Let $a_n = 1$.

$$f(x) = (x - 1)(x - 3i)(x + 3i)$$
$$= (x - 1)(x^2 + 9)$$
$$= \boxed{} + 9x - x^2 - 9$$
$$= x^3 - \boxed{} + 9x - 9$$

Example 2   Find a polynomial function of degree 5 with -1 as a zero of multiplicity 3, 4 as a zero of multiplicity 1, and 0 as a zero of multiplicity 1.

If these are the zeros, then the function in factored form must be:

$$f(x) = a_n \left(x + \boxed{}\right)^3 \left(x - \boxed{}\right)(x - 0)$$

Let $a_n = 1$.

$$f(x) = (x+1)^3 (x-4)(x-0)$$
$$= x^5 - \boxed{} - 9x^3 - \boxed{} - 4x$$

Nonreal Zeros: $a + bi$ and $a - bi,\ b \neq 0$

If a complex number $a + bi$, $b \neq 0$, is a zero of a polynomial function $f(x)$ with *real* coefficients, then its conjugate, $a - bi$, is also a zero. For example, if $2 + 7i$ is a zero of a polynomial function $f(x)$, with real coefficients, then its conjugate, $2 - 7i$, is also a zero. (Nonreal zeros occur in conjugate pairs.)

Irrational Zeros: $a + c\sqrt{b}$ and $a - c\sqrt{b}$, b Is Not a Perfect Square

If $a + c\sqrt{b}$, where a, b, and c are rational and b is not a perfect square, is a zero of a polynomial function $f(x)$ with *rational* coefficients, then its conjugate, $a - c\sqrt{b}$, is also a zero. For example, if $-3 + 5\sqrt{2}$ is a zero of a polynomial function $f(x)$ with rational coefficients, then its conjugate, $-3 - 5\sqrt{2}$, is also a zero. (Irrational zeros occur in conjugate pairs.)

Example 3   Suppose that a polynomial function of degree 6 with rational coefficients has $-2 + 5i$, $-2i$, and $1 - \sqrt{3}$ as three of its zeros. Find the other zeros.

Since the function has rational coefficients, the other zeros are the conjugates of the given zeros. They are

$$-2 - 5i,\ \boxed{},\ \text{and } 1 + \sqrt{3}.$$

Example 4 Find a polynomial function of lowest degree with rational coefficients that has $-\sqrt{3}$, and $1+i$ as two of its zeros.

Two other zeros must be ⬚ and ⬚. (We will not include additional zeros, because we want the function of lowest degree.)

$$f(x) = \left[x - \left(-\sqrt{3}\right)\right]\left[x - \left(\boxed{}\right)\right]\left[x - (1+i)\right]\left[x - \left(\boxed{}\right)\right]$$

$$= \left(x + \sqrt{3}\right)\left(x - \sqrt{3}\right)\left((x-1) - i\right)\left((x-1) + \boxed{}\right)$$

$$= \left(x^2 - \boxed{}\right)\left((x-1)^2 - i^2\right)$$

$$= \left(x^2 - 3\right)\left(x^2 - 2x + 1^2 + 1\right)$$

$$= \left(x^2 - 3\right)\left(x^2 - 2x + 2\right)$$

$$= x^4 - 2x^3 - \boxed{} + 6x - \boxed{}$$

The Rational Zeros Theorem

Let $P(x) = a_n x^n + a_{n-1} x^{n-1} + \ldots + a_1 x + a_0$, where all the coefficients are integers. Consider a rational number denoted by p/q, where p and q are relatively prime (having no common factor besides -1 and 1). If p/q is a zero of $P(x)$, then p is a factor of a_0 and q is a factor of a_n.

Example 5   Given $f(x) = 3x^4 - 11x^3 + 10x - 4$:

a) Find the rational zeros and then the other zeros; that is, solve $f(x) = 0$.

b) Factor $f(x)$ into linear factors.

First, notice the degree of the polynomial is 4, so there are at most 4 distinct zeros.

a) The rational zeros theorem says that if a rational number p/q is a zero of $f(x)$, then p must be a factor of -4 and q must be a factor of 3.

$$\frac{\text{Possibilities for } p}{\text{Possibilities for } q} : \frac{\pm 1, \ \pm 2, \ \pm 4}{\pm 1, \ \pm 3}$$

$$\text{Possibilities for } \frac{p}{q} : 1, -1, 2, -2, 4, -4, \frac{1}{3}, -\frac{1}{3}, \frac{2}{3}, -\frac{2}{3}, \frac{4}{3}, -\frac{4}{3}$$

To find which are zeros, we use synthetic division.

Try 1:

$$
\begin{array}{r|rrrrr}
1 & 3 & -11 & 0 & 10 & -4 \\
  &   & 3 & \Box & -8 & 2 \\
\hline
  & \Box & -8 & -8 & 2 & \Box
\end{array}
$$

Since $f(1) = \Box$, 1 is not a zero.

Try -1:

$$
\begin{array}{r|rrrrr}
-1 & 3 & -11 & 0 & 10 & -4 \\
   &   & \Box & 14 & -14 & \Box \\
\hline
   & 3 & -14 & \Box & -4 & 0
\end{array}
$$

Since $f(-1) = \Box$, -1 is a zero.

Using the results of the synthetic division, we can rewrite $f(x)$ as

$$f(x) = (x+1)(3x^3 - 14x^2 + 14x - 4).$$

Now consider $3x^3 - 14x^2 + 14x - 4$. Continue using synthetic division to find other zeros.

Try -1 again since it might have multiplicity 2.

$$
\begin{array}{r|rrrr}
-1 & 3 & -14 & 14 & -4 \\
   &   & \Box & 17 & \Box \\
\hline
   & 3 & -17 & \Box & -35
\end{array}
$$

Since $f(-1) = -35$, -1 does not have multiplicity 2.

Synthetic division shows that 2, -2, 4, and -4 are not zeros.

Now try $\dfrac{2}{3}$:

$$
\begin{array}{r|rrrr}
2/3 & 3 & -14 & 14 & -4 \\
    &   & 2 & \Box & 4 \\
\hline
    & \Box & -12 & 6 & \vert\ \Box
\end{array}
$$

Since $f\left(\dfrac{2}{3}\right) = \boxed{}$, $\dfrac{2}{3}$ is a zero of $f(x)$ and $\left(x - \dfrac{2}{3}\right)$ is a factor of $f(x)$.

We can rewrite $f(x)$ as

$$f(x) = (x+1)\left(x - \frac{2}{3}\right)\left(3x^2 - 12x + 6\right)$$

$$= (x+1)\left(x - \frac{2}{3}\right)3\left(x^2 - 4x + 2\right)$$

and the last factor can be used to find the two remaining zeros by solving the equation $x^2 - 4x + 2 = 0$ using the quadratic formula.

$$x = \frac{-b \pm \sqrt{b^2 - 4ac}}{2a}$$

$$x = \frac{-(-4) \pm \sqrt{(-4)^2 - 4 \cdot 1 \cdot 2}}{2 \cdot 1}$$

$$x = \frac{4 \pm \sqrt{\boxed{}}}{2} = \frac{4 \pm 2\sqrt{2}}{2} = \frac{2\left(2 \pm \sqrt{2}\right)}{2} = 2 \pm \sqrt{2}$$

So the rational zeros are $\boxed{}$, $\dfrac{2}{3}$, $2 + \sqrt{2}$, $2 - \sqrt{2}$.

b) The complete factorization of $f(x)$ is

$$f(x) = 3(x+1)\left(x - \frac{2}{3}\right)\left[x - \left(2 - \sqrt{2}\right)\right]\left[x - \left(2 + \sqrt{2}\right)\right], \text{ or}$$

$$f(x) = (x+1)\left(3x - \boxed{}\right)\left[x - \left(2 - \sqrt{2}\right)\right]\left[x - \left(2 + \sqrt{2}\right)\right].$$

Example 6   Given $f(x) = 2x^5 - x^4 - 4x^3 + 2x^2 - 30x + 15$:

a) Find the rational zeros and then the other zeros; that is, solve $f(x) = 0$.

b) Factor $f(x)$ into linear factors.

First, notice the degree of the polynomial is 5, so there are at most 5 distinct zeros.

a) The rational zeros theorem says that if a rational number p/q is a zero of $f(x)$, then p must be a factor of 15 and q must be a factor of 2.

$$\frac{\textit{Possibilities for } p}{\textit{Possibilities for } q} : \frac{\pm 1, \ \pm 3, \ \pm 5, \ \pm 15}{\pm 1, \ \pm 2}$$

$\textit{Possibilities for } \dfrac{p}{q} : 1, -1, 3, -3, 5, -5, 15, -15, \dfrac{1}{2}, -\dfrac{1}{2}, \dfrac{3}{2}, -\dfrac{3}{2}, \dfrac{5}{2}, -\dfrac{5}{2}, \dfrac{15}{2}, -\dfrac{15}{2}$

To find which are zeros, we use synthetic division.

Try 1:

1⌋	2	−1	−4	2	−30	15
		2	1	−3	−1	−31
	2	1	−3	−1	−31	☐

Since $f(1) = \boxed{}$, 1 is not a zero.

Try −1:

−1⌋	2	−1	−4	2	−30	15
		−2	3	1	−3	33
	2	−3	−1	3	−33	☐

Since $f(-1) = \boxed{}$, −1 is not a zero.

Try $\dfrac{1}{2}$:

1/2⌋	2	−1	−4	2	−30	15
		1	0	−2	0	−15
	2	☐	−4	0	−30	☐

Since $f\left(\dfrac{1}{2}\right) = \boxed{}$, $\dfrac{1}{2}$ is a zero of $f(x)$ and $\left(x - \dfrac{1}{2}\right)$ is a factor of $f(x)$.

We can rewrite $f(x)$ as

$$f(x) = \left(x - \frac{1}{2}\right)\left(2x^4 - 4x^2 - 30\right)$$

$$= \left(x - \frac{1}{2}\right)2\left(x^4 - 2x^2 - 15\right) = \left(x - \frac{1}{2}\right)2\left(x^2 - 5\right)\left(x^2 + 3\right).$$

We now solve the equation $f(x) = 0$ using the principle of zero products.

$$\left(x - \frac{1}{2}\right)2\left(x^2 - 5\right)\left(x^2 + 3\right) = 0$$

$x - \dfrac{1}{2} = 0 \qquad$ or $\quad x^2 - 5 = 0 \qquad$ or $\quad x^2 + 3 = 0$

$x = \dfrac{1}{2} \qquad$ or $\qquad x^2 = 5 \qquad$ or $\qquad x^2 = -3$

$x = \boxed{} \qquad$ or $\qquad x = \boxed{} \qquad$ or $\qquad x = \boxed{}$

The rational zero is $\dfrac{1}{2}$. The other zeros are $\sqrt{5}, \ -\sqrt{5}, \ \sqrt{3}i,$ and $-\sqrt{3}i$.

b) The complete factorization of $f(x)$ is

$$f(x) = 2\left(x - \boxed{}\right)\left(x + \sqrt{5}\right)\left(x - \sqrt{5}\right)\left(x + \sqrt{3}i\right)\left(x - \boxed{}\right), \text{ or}$$

$$f(x) = (2x - 1)\left(x + \sqrt{5}\right)\left(x - \boxed{}\right)\left(x + \sqrt{3}i\right)\left(x - \sqrt{3}i\right).$$

Descartes' Rule of Signs

Let $P(x)$, written in descending order or ascending order, be a polynomial function with real coefficients and a nonzero constant term. The number of positive real zeros of $P(x)$ is either:

1. The same as the number of variations of sign in $P(x)$, or

2. Less than the number of variations of sign in $P(x)$ by a positive even integer.

The number of negative real zeros of $P(x)$ is either:

3. The same as the number of variations of sign in $P(-x)$, or

4. Less than the number of variations of sign in $P(-x)$ by a positive even integer.

A zero of multiplicity m must be counted m times.

Example 7 Determine the number of variations of sign in the polynomial function $P(x) = 2x^5 - 3x^2 + x + 4$.

There are $\boxed{}$ variations of sign.

Example 8 Given the function $P(x) = 2x^5 - 5x^2 - 3x + 6$, what does Descartes' rule of signs tell you about the number of positive real zeros and the number of negative real zeros?

There are $\boxed{}$ variations of sign in $P(x)$.

Thus the number of positive real zeros is $\boxed{}$ or $2 - 2 = \boxed{}$.

Now find $P(-x)$:

$$P(-x) = 2(-x)^5 - 5(-x)^2 - 3(-x) + 6$$

$$= -2x^5 \boxed{} 5x^2 + 3x + 6$$

There is $\boxed{}$ variation of sign in $P(-x)$.

Thus there is $\boxed{}$ negative real zero.

Total Number of Zeros	5	
Positive Real	2	0
Negative Real	1	1
Nonreal	2	4

Example 9   Given the function $P(x) = 5x^4 - 3x^3 + 7x^2 - 12x + 4$, what does Descartes' rule of signs tell you about the number of positive real zeros and the number of negative real zeros?

There are $\boxed{}$ variations of sign in $P(x)$.

Thus $P(x)$ has $\boxed{}$ or $4 - 2 = \boxed{}$ or $4 - 4 = \boxed{}$ positive real zeros.

Now find $P(-x)$:

$$P(-x) = 5(-x)^4 - 3(-x)^3 + 7(-x)^2 - 12(-x) + 4$$

$$= 5x^4 + 3x^3 + 7x^2 + 12x + 4$$

There are $\boxed{}$ variations of sign in $P(-x)$.

Thus there are $\boxed{}$ negative real zeros.

Total Number of Zeros	4		
Positive Real	4	2	0
Negative Real	0	0	0
Nonreal	0	2	4

Example 10   Given the function $P(x) = 6x^6 - 2x^2 - 5x$, what does Descartes' rule of signs tell you about the number of positive real zeros and the number of negative real zeros?

$$P(x) = 6x^6 - 2x^2 - 5x = x(6x^5 - 2x - 5)$$

Note that $x = 0$ is a zero of this function.

Let $Q(x) = 6x^5 - 2x - 5$.

There is $\boxed{}$ variation of sign in $Q(x)$.

Thus there is $\boxed{}$ positive real zero.

Now find $Q(-x)$:

$$Q(-x) = 6(-x)^5 - 2(-x) - 5 = -6x^5 + 2x - 5$$

There are $\boxed{}$ variations of sign in $Q(-x)$.

Thus there are $\boxed{}$ or $2 - 2 = \boxed{}$ negative real zeros.

Total Number of Zeros	6	
0 as a Zero	1	1
Positive Real	1	1
Negative Real	2	0
Nonreal	2	4

Section 4.5 Rational Functions

Rational Function

A **rational function** is a function f that is a quotient of two polynomials. That is,

$$f(x) = \frac{p(x)}{q(x)}$$

where $p(x)$ and $q(x)$ are polynomials and where $q(x)$ is not the zero polynomial. The domain of f consists of all inputs x for which $q(x) \neq 0$.

Example 1   Consider $f(x) = \dfrac{1}{x-3}$. Find the domain and graph f.

The domain is all the x-values which do not make the denominator zero.

$x - 3 = 0$ when $x = \boxed{}$.

Domain: $\left\{x \mid x \neq \boxed{}\right\}$, or $(-\infty, 3) \cup (3, \infty)$

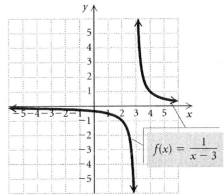

$$f(x) = \frac{1}{x-3}$$

Example 2   Determine the domain of each of the functions.

Function	Domain
$f(x) = \dfrac{1}{x}$	$\left\{x \mid x \neq \boxed{}\right\}$, or $(-\infty, 0) \cup (0, \infty)$
$f(x) = \dfrac{1}{x^2}$	$\{x \mid x \neq 0\}$, or $\left(-\infty, \boxed{}\right) \cup (0, \infty)$
$f(x) = \dfrac{x-3}{x^2 + x - 2} = \dfrac{x-3}{(x+2)(x-1)}$	$\left\{x \mid x \neq -2 \text{ and } x \neq \boxed{}\right\}$, or $(-\infty, -2) \cup (-2, 1) \cup (1, \infty)$
$f(x) = \dfrac{2x+5}{2x-6} = \dfrac{2x+5}{2(x-3)}$	$\left\{x \mid x \neq \boxed{}\right\}$, or $(-\infty, 3) \cup (3, \infty)$
$f(x) = \dfrac{x^2 + 2x - 3}{x^2 - x - 2} = \dfrac{x^2 + 2x - 3}{(x+1)(x-2)}$	$\{x \mid x \neq -1 \text{ and } x \neq 2\}$, or $(-\infty, -1) \cup (-1, 2) \cup (2, \infty)$
$f(x) = \dfrac{-x^2}{x+1}$	$\{x \mid x \neq -1\}$, or $(-\infty, -1) \cup \left(\boxed{}, \infty\right)$

> **Determining Vertical Asymptotes**
>
> For a rational function $f(x) = p(x)/q(x)$, where $p(x)$ and $q(x)$ are polynomials with *no common factors other than constants*, if a is a zero of the denominator, then the line $x = a$ is a vertical asymptote for the graph of the function.

Example 3   Determine the vertical asymptotes for the graph of each of the following functions.

a) $f(x) = \dfrac{2x - 11}{x^2 + 2x - 8}$

First, factor the denominator.

$$f(x) = \frac{2x - 11}{(x + 4)\left(x - \boxed{}\right)}$$

The numerator and the denominator have no common factors.

Zeros: $-4,\ \boxed{}$

Vertical asymptotes: $x = \boxed{}$ and $x = 2$

b) $h(x) = \dfrac{x^2 - 4x}{x^3 - x}$

First, factor the numerator and the denominator.

$$h(x) = \frac{x\left(x - \boxed{}\right)}{x\left(x^2 - 1\right)} = \frac{x\left(x - \boxed{}\right)}{x(x + 1)(x - 1)}$$

Zeros of the numerator: $0,\ \boxed{}$

Zeros of the denominator: $0,\ -1,$ and $\boxed{}$

Vertical asymptotes: $x = -1$ and $x = 1$

c) $g(x) = \dfrac{x - 2}{x^3 - 5x}$

First, factor the denominator: $g(x) = \dfrac{x - 2}{x\left(x^2 - 5\right)}$.

The numerator and the denominator have no common factors.

Solving $x\left(x^2 - 5\right) = 0$, we have $x = \boxed{}$ or $x = \pm\sqrt{5}$.

Zeros of the denominator: $0,\ -\sqrt{5},$ and $\boxed{}$

Vertical asymptotes: $x = \boxed{}$, $x = -\sqrt{5}$, and $x = \sqrt{5}$

Example 4   Find the horizontal asymptote: $f(x) = \dfrac{-7x^4 - 10x^2 + 1}{11x^4 + x - 2}$.

When the numerator and denominator of a rational function have the same degree, the horizontal asymptote is given by $y = a/b$ where a and b are the leading coefficients of the numerator and the denominator, respectively.

Thus, the horizontal asymptote is $y = \dfrac{-7}{\Box} = -\dfrac{\Box}{11}$, or $-0.\overline{63}$.

Example 5   Find the horizontal asymptote: $f(x) = \dfrac{2x + 3}{x^3 - 2x^2 + 4}$.

Let's call the numerator $p(x)$ and the denominator $q(x)$. Notice $p(x)$ has degree 1 and $q(x)$ has degree 3. As x increases, a degree 3 polynomial will increase much faster than a degree 1 polynomial.

So, as $x \to \infty$, $\dfrac{p(x)}{q(x)} \to 0$, and as $x \to -\infty$, $\dfrac{p(x)}{q(x)} \to 0$.

Thus, the horizontal asymptote is $y = \boxed{}$.

This is the case anytime the degree of the numerator is less than the degree of the denominator.

Determining a Horizontal Asymptote

- When the numerator and the denominator of a rational function have the same degree, the line $y = a/b$ is the horizontal asymptote, where a and b are the leading coefficients of the numerator and the denominator, respectively.

- When the degree of the numerator of a rational function is less than the degree of the denominator, the x-axis, or $y = 0$, is the horizontal asymptote.

- When the degree of the numerator of a rational function is greater than the degree of the denominator, there is no horizontal asymptote.

Example 6   Graph $g(x) = \dfrac{2x^2 + 1}{x^2}$. Include and label all asymptotes.

The denominator is zero when $x = \boxed{}$, so the vertical asymptote is $x = 0$.

The degree of the numerator is the same as the degree of the denominator, so the ratio of the leading coefficients gives us the horizontal asymptote: $y = \boxed{}$.

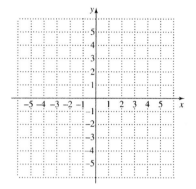

Find and plot some points on the graph. We have:

$$(-2, 2.25), \ (-1, 3), \ (-0.5, 6),$$
$$(0.5, 6), \ (1.5, 2.\overline{4}), \ (2, 2.25)$$

Sometimes a line that is neither horizontal nor vertical is an asymptote. Such a line is called an **oblique asymptote**, or a **slant asymptote**.

Example 7   Find all the asymptotes of $f(x) = \dfrac{2x^2 - 3x - 1}{x - 2}$.

The denominator is zero when $x = \boxed{}$, so the vertical asymptote is $x = 2$.

There are no horizontal asymptotes, because the degree of the numerator is greater than that of the denominator.

When the degree of the numerator is 1 more than the degree of the denominator, we divide to find an equivalent expression.

$$f(x) = \frac{2x^2 - 3x - 1}{x - 2} = 2x + 1 + \frac{1}{x - 2}$$

As $x \to \infty$ or as $x \to -\infty$, $\dfrac{1}{x - 2} \to 0$, so $f(x) \to 2x + 1$.

The line $y = \boxed{}$ is an oblique asymptote.

Example 8   Graph: $f(x) = \dfrac{2x+3}{3x^2 + 7x - 6}$.

First, find the zeros of the denominator.

$$3x^2 + 7x - 6 = 0$$

$$(3x - 2)(x + 3) = 0$$

$$3x - 2 = 0 \quad or \quad x + 3 = 0$$

$$x = \frac{2}{3} \quad or \quad x = \boxed{}$$

Neither zero is a zero of the numerator, so the vertical asymptotes are $x = \dfrac{2}{3}$ and $x = -3$.

The degree of the numerator is less than the degree of the denominator, so the horizontal asymptote is $y = \boxed{}$.

The zero of the numerator tells us the x-intercept.

$$2x + 3 = 0$$

$$2x = -3$$

$$x = \boxed{}$$

Thus the x-intercept is $\left(-\dfrac{3}{2}, 0\right)$.

To find the y-intercept, find $f(0)$.

$$f(0) = \frac{2 \cdot 0 + 3}{3 \cdot 0^2 + 7 \cdot 0 - 6}$$

$$= \frac{3}{-6} = -\frac{1}{2}$$

The y-intercept is $\left(0, -\dfrac{1}{2}\right)$.

Finally, choose some other x-values and find their corresponding function values to find enough points to determine the shape of the graph.

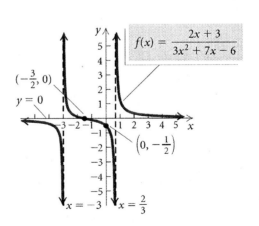

Example 9   Graph: $g(x) = \dfrac{x^2 - 1}{x^2 + x - 6}$.

First, find the zeros of the denominator.

$$x^2 + x - 6 = 0$$
$$(x + 3)(x - 2) = 0$$
$$x = -3 \quad or \quad x = 2$$

Neither zero is a zero of the numerator, so the vertical asymptotes are $x = -3$ and $x = \boxed{}$.

The degree of the numerator is the same as the degree of the denominator, so the horizontal asymptote is $y = \dfrac{1}{1} = 1$.

The zeros of the numerator tell us the *x*-intercepts.

$$x^2 - 1 = 0$$
$$(x + 1)(x - 1) = 0$$
$$x = -1 \quad or \quad x = \boxed{}$$

The *x*-intercepts are $(-1, 0)$ and $(1, 0)$.

To find the *y*-intercept, find $g(0)$.

$$g(0) = \frac{0^2 - 1}{0^2 + 0 - 6} = \frac{-1}{-6} = \frac{1}{6}$$

The *y*-intercept is $\left(0, \dfrac{1}{6}\right)$.

Finally, choose some other *x*-values and find their corresponding function values to find enough points to determine the shape of the graph.

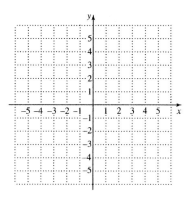

Example 10   Graph: $g(x) = \dfrac{x - 2}{x^2 - x - 2}$.

Factoring the denominator, we have

$$g(x) = \frac{x - 2}{(x - 2)(x + 1)}.$$

The denominator is equal to 0 when $x = 2$ or $x = -1$.

Domain: $\{x \mid x \neq -1 \text{ and } x \neq 2\}$

2 is also a zero of the numerator, so the only vertical asymptote is $x = -1$.

The degree of the numerator is less than the degree of the denominator, so the horizontal asymptote is $y = \boxed{}$.

The function has no zeros because the only zero of the numerator, 2, is not in the domain of the function.

$$g(x) = \frac{x-2}{x^2-x-2} = \frac{x-2}{(x-2)(x+1)} = \frac{1}{x+1}, \ x \neq 2$$

The graph will be the graph of

$g(x) = \dfrac{1}{x+1}$ with a hole at $x = 2$.

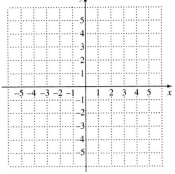

Example 11   Graph: $f(x) = \dfrac{-2x^2-x+15}{x^2-x-12}$.

Factoring: $f(x) = \dfrac{(-2x+5)(x+3)}{(x-4)(x+3)}$

Domain: $\left\{ x \big| x \neq \boxed{} \ \text{and} \ x \neq \boxed{} \right\}$, or $(-\infty, -3) \cup (-3, 4) \cup (4, \infty)$

The only zero of the denominator that is not a zero of the numerator is 4, so the only vertical asymptote is $x = 4$.

The degree of the numerator is equal to the degree of the denominator, so the horizontal asymptote is $y = \dfrac{-2}{1} = -2$.

The zeros of the numerator are $\dfrac{5}{2}$ and -3, but -3 is not in the domain of the function, so the only x-intercept is $\left(\dfrac{5}{2}, 0 \right)$.

To find the y-intercept, find $f(0)$.

$$f(0) = \frac{15}{-12} = -\frac{5}{4}$$

The y-intercept is $\left(0, -\dfrac{5}{4} \right)$.

Simplifying: $f(x) = \dfrac{(-2x+5)(x+3)}{(x-4)(x+3)} = \dfrac{-2x+5}{x-4}$, $x \neq -3$ and $x \neq 4$

The graph will be the graph of $f(x) = \dfrac{-2x+5}{x-4}$ with a hole at $x = -3$.

Determine the second coordinate of the hole:

$$f(-3) = \frac{-2(-3)+5}{-3-4} = \frac{11}{\boxed{}} = -\frac{11}{7}$$

The hole is at $\left(-3, \boxed{}\right)$, indicated by an open circle on the graph.

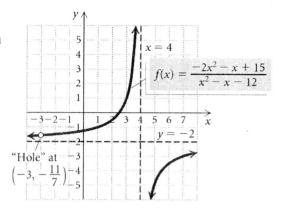

$$f(x) = \frac{-2x^2 - x + 15}{x^2 - x - 12}$$

"Hole" at $\left(-3, -\frac{11}{7}\right)$

Example 12 *Temperature During an Illness.* A person's temperature T, in degrees Fahrenheit, during an illness is given by the function

$$T(t) = \frac{4t}{t^2 + 1} + 98.6,$$

where time t is given in hours since the onset of the illness.

a) Find the temperature at $t = 0, 1, 2, 5, 12,$ and 24.

$$T(0) = 98.6$$

$$T(1) = 100.6$$

$$T(2) = \boxed{}$$

$$T(5) = 99.369$$

$$T(12) = 98.931$$

$$T(24) = \boxed{}$$

b) Find the horizontal asymptote of the graph of $T(t)$. Complete $T(t) \rightarrow \boxed{}$ as $t \rightarrow \infty$.

$$T(t) = \frac{4t}{t^2 + 1} + 98.6$$

$$= \frac{4t + 98.6t^2 + 98.6}{t^2 + 1} \qquad \text{Writing with a common denominator}$$

$$= \frac{98.6t^2 + 4t + 98.6}{t^2 + 1}$$

Since the degree of the numerator is the same as the degree of the denominator, the horizontal asymptote is the ratio of the leading coefficients: $y = \dfrac{98.6}{1}$, or $y = \boxed{}$.

So, $T(t) \rightarrow \boxed{}$ as $t \rightarrow \infty$.

c) Give the meaning of the answer to part (b) in terms of the application.
As time goes on, a person's temperature returns to normal, $98.6°\,$F.

Section 4.6   Polynomial Inequalities and Rational Inequalities

Example 1   Solve: $x^2 - 4x - 5 > 0$.

The related equation is $x^2 - 4x - 5 = 0$.

Find the x-intercepts by solving the related equation.

Graph $f(x) = x^2 - 4x - 5$.

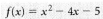

$$x^2 - 4x - 5 = 0$$

$$(x - 5)(x + 1) = 0$$

$$x - 5 = 0 \quad \text{or} \quad x + 1 = 0$$

$$x = \boxed{} \quad \text{or} \quad x = \boxed{}$$

x-intercepts: $(5, 0)$ and $(-1, 0)$

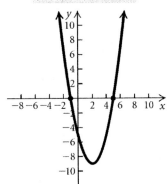

Interval	$(-\infty, -1)$	$(-1, 5)$	$(5, \infty)$
Test value	$f(-2) = 7$	$f(0) = -5$	$f(7) = 16$
Sign of $f(x)$	$+$	$-$	$+$

The solution set is $\left(-\infty, \boxed{}\right) \cup \left(\boxed{}, \infty\right)$, *or* $\left\{x \middle| x < -1 \text{ or } x \boxed{} 5\right\}$.

Example 2   Solve: $x^2 + 3x - 5 \le x + 3$.

$$x^2 + 3x - 5 - x - 3 \le 0$$

$$x^2 + 2x - 8 \le 0$$

Graph $f(x) = x^2 + 2x - 8$.

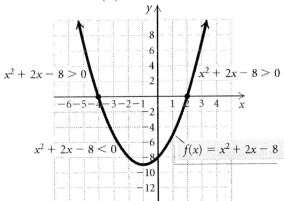

Find the x-intercepts by solving the related equation.

$$x^2 + 2x - 8 = 0$$

$$\left(x + \boxed{}\right)\left(x - \boxed{}\right) = 0$$

$$x + 4 = 0 \quad \text{or} \quad x - 2 = 0$$

$$x = -4 \quad \text{or} \quad x = 2$$

x-intercepts: $\left(\boxed{}, 0\right),\ \left(\boxed{}, 0\right)$

The solution set is $\left[-4, \boxed{}\right]$, *or* $\left\{x \middle| \boxed{} \le x \le 2\right\}$.

Example 3   Solve: $x^3 - x > 0$.

Find the zeros of the related equation.

$$x^3 - x = 0$$

$$x\left(x^2 - \boxed{}\right) = 0$$

$$x(x+1)(x-1) = 0$$

$x = 0$ *or* $x + 1 = 0$ *or* $\boxed{} = 0$

$x = 0$ *or* $x = \boxed{}$ *or* $x = 1$

x-intercepts: $(0,0)$, $(-1,0)$, $\left(\boxed{},0\right)$

Interval	$(-\infty,-1)$	$(-1,0)$	$(0,1)$	$(1,\infty)$
Test value	$f(-2) = -6$	$f(-0.5) = 0.375$	$f(0.5) = -0.375$	$f(2) = 6$
Sign of $f(x)$	$-$	$+$	$-$	$+$

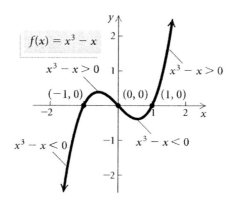

The solution set is $\left(-1, \boxed{}\right) \cup \left(\boxed{}, \infty\right)$, or $\left\{x \mid -1 < x < 0 \text{ or } x > \boxed{}\right\}$.

Example 4   Solve: $3x^4 + 10x \le 11x^3 + 4$.

$$3x^4 - 11x^3 + 10x - 4 \le 0$$

Find the zeros of the related equation $3x^4 - 11x^3 + 10x - 4 = 0$.

This equation was solved in Example 5 in Section 4.4. The exact solutions are

$$-1,\ 2 - \sqrt{2},\ \frac{2}{3},\ \text{and } 2 + \sqrt{2}.$$

The approximate solutions are

$$-1,\ 0.586,\ 0.667,\ \text{and } 3.414.$$

Interval	Test Value	Sign of $f(x)$
$(-\infty, \ -1)$	$f(-2) = 112$	+
$\left(-1, \ 2-\sqrt{2}\right)$	$f(0) = -4$	−
$\left(2-\sqrt{2}, \ \dfrac{2}{3}\right)$	$f(0.6) = 0.0128$	+
$\left(\dfrac{2}{3}, \ 2+\sqrt{2}\right)$	$f(1) = -2$	−
$\left(2+\sqrt{2}, \ \infty\right)$	$f(4) = 100$	+

The solution set is $\left[\boxed{}, 2-\sqrt{2}\right] \cup \left[\dfrac{2}{3}, 2+\sqrt{2}\right]$, or

$\left\{ x \middle| -1 \le x \le 2-\sqrt{2} \ or \ \dfrac{2}{3} \le x \le \boxed{} \right\}.$

Example 5   Solve: $\dfrac{3x}{x+6} < 0.$

Related equation in function form:

$$f(x) = \dfrac{3x}{x+6}$$

To find the critical values, first find the values of x for which the function is not defined.

$x+6 = 0$ when $x = \boxed{}.$

The values of x for which the related function equals 0 are also critical values.

$$f(x) = 0$$

$$\dfrac{3x}{x+6} = 0$$

$$\left(x + \boxed{}\right) \cdot \dfrac{3x}{x+6} = 0 \cdot \left(x + \boxed{}\right)$$

$$3x = 0$$

$$x = \boxed{}$$

Interval	$(-\infty, -6)$	$(-6, 0)$	$(0, \infty)$
Test value	$f(-8) = 12$	$f(-2) = -\dfrac{3}{2}$	$f(3) = 1$
Sign of $f(x)$	+	−	+

The solution set is $\left(-6, \boxed{}\right)$, or $\left\{ x \middle| \boxed{} < x < 0 \right\}.$

Example 6   Solve: $\dfrac{x+1}{2x-4} \leq 1$.

$$\dfrac{x+1}{2x-4} - 1 \leq 0$$

Related equation in function form:

$$f(x) = \dfrac{x+1}{2x-4} - 1$$

Critical values are the values of x for which $f(x)$ is not defined or 0. $f(x)$ is not defined when $2x - 4 = 0$, or $x = 2$.

Now solve $f(x) = 0$.

$$\dfrac{x+1}{2x-4} - 1 = 0$$

$$\dfrac{x+1}{2x-4} = 1$$

$$x + 1 = 2x - 4$$

$$1 = x - 4$$

$$\boxed{} = x$$

Interval	Test Value	Sign of $f(x)$
$(-\infty,\ 2)$	$f(0) = -\dfrac{5}{4}$	$\boxed{}$
$(2,\ 5)$	$f(3) = 1$	$\boxed{}$
$(5,\ \infty)$	$f(6) = -\dfrac{1}{8}$	$\boxed{}$

The solution set is $\left(-\infty, \boxed{}\right) \cup \left[\boxed{}, \infty\right)$.

Example 7 Solve: $\dfrac{x-3}{x+4} \ge \dfrac{x+2}{x-5}$.

$$\frac{x-3}{x+4} - \frac{x+2}{x-5} \ge 0$$

Related equation in function form:

$$f(x) = \frac{x-3}{x+4} - \frac{x+2}{x-5}$$

Find the critical values.

$f(x)$ is not defined for $x = -4$ and $x = \boxed{}$.

Now solve $f(x) = 0$.

$$\frac{x-3}{x+4} - \frac{x+2}{x-5} = 0$$

$$(x+4)\cdot(x-5)\cdot\left[\frac{x-3}{x+4} - \frac{x+2}{x-5}\right] = 0\cdot(x+4)\cdot(x-5)$$

$$(x-5)\cdot\left(x-\boxed{}\right) - (x+4)\cdot\left(x+\boxed{}\right) = 0$$

$$x^2 - 8x + 15 - \left(x^2 + \boxed{}x + 8\right) = 0$$

$$x^2 - 8x + 15 - x^2 - 6x - 8 = 0$$

$$-14x + 7 = 0$$

$$-14x = -7$$

$$x = \frac{7}{14} = \boxed{}$$

Critical values: $\boxed{}$, $\dfrac{1}{2}$, and $\boxed{}$

Interval	Test Value	Sign of $f(x)$
$(-\infty,\ -4)$	$f(-5) = 7.7$	$+$
$\left(-4,\ \dfrac{1}{2}\right)$	$f(-2) = -2.5$	$\boxed{}$
$\left(\dfrac{1}{2},\ 5\right)$	$f(3) = 2.5$	$\boxed{}$
$(5,\ \infty)$	$f(6) = -7.7$	$-$

The critical values -4 and 5 are not in the domain. The critical value $\dfrac{1}{2}$ is.

The solution set is $\left(-\infty, \boxed{}\right) \cup \left[\dfrac{1}{2}, 5\right)$, or $\left\{ x \mid x < \boxed{} \ or \ \dfrac{1}{2} \le x < \boxed{} \right\}$.

(handwritten) $x = y^2 - 5$ $y = \sqrt{x+5}$

(handwritten) $\sqrt{y^2} = x + 5$

Section 5.1   Inverse Functions

Inverse Relation

Interchanging the first and second coordinates of each ordered pair in a relation produces the **inverse relation**.

Example 1   Consider the relation g given by
$$g = \{(2, 4), (-1, 3), (-2, 0)\}.$$

Graph the relation in blue. Find the inverse relation and graph it in red.

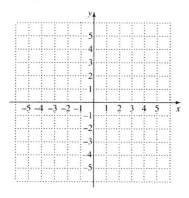

Inverse of $g = \{ \boxed{} \}$

Inverse Relation

If a relation is defined by an equation, interchanging the variables produces an equation of the inverse relation.

Example 2   Find an equation for the inverse of the relation: $y = x^2 - 5x$.

The equation of the inverse relation is $\boxed{}$.

One-to-One Functions

A function f is **one-to-one** if different inputs have different outputs – that is,

if $a \neq b$, then $f(a) \neq f(b)$.

Or a function is **one-to-one** if when the outputs are the same, the inputs are the same – that is,

if $f(a) = f(b)$, then $a = b$.

If the inverse of a function f is also a function, it is named f^{-1} (read "f-inverse").

One-to-One Functions and Inverses
- If a function f is one-to-one, then its inverse f^{-1} is a function.
- The domain of a one-to-one function f is the range of the inverse f^{-1}.
- The range of a one-to-one function f is the domain of the inverse f^{-1}.
- A function that is increasing over its entire domain or is decreasing over its entire domain is a one-to-one function.

Example 3   Given the function f described by $f(x) = 2x - 3$, prove that f is one-to-one (that is, it has an inverse that is a function).

Given $f(x) = 2x - 3$.

If $f(a) = f(b)$, then $\boxed{}$.

$f(a) = \boxed{}$ $f(b) = \boxed{}$

$2a - 3 = 2b - 3$

$\boxed{} = \boxed{}$

$\boxed{} = \boxed{}$

Thus, if $f(a) = f(b)$, then $a = b$.

This tells us that $f(x)$ is one-to-one.

Example 4   Given the function g described by $g(x) = x^2$, prove that g is not one-to-one.

To show that g is not a one-to-one function, we need to find two numbers such that $a \neq b$ and $g(a) = g(b)$.

Try the values $x = 3$ and $x = -3$.

$g\left(\boxed{}\right) = 3^2 = \boxed{}$

$g\left(\boxed{}\right) = (-3)^2 = \boxed{}$

Thus, g _____ one-to-one.
        is / is not

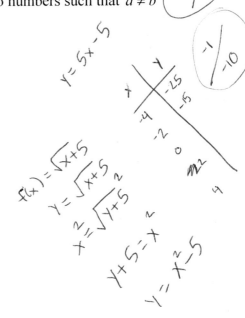

Horizontal-Line Test

If it is possible for a horizontal line to intersect the graph of a function more than once, then the function is *not* one-to-one and its inverse is *not* a function.

Example 5   From the graphs shown, determine whether each function is one-to-one and thus has an inverse that is a function.

a)

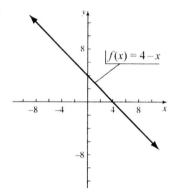

one-to-one / not one-to-one

b)

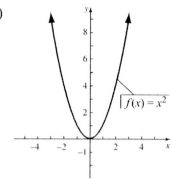

one-to-one / not one-to-one

c)

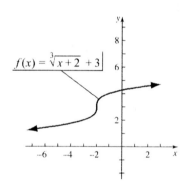

one-to-one / not one-to-one

d)

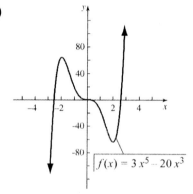

one-to-one / not one-to-one

Obtaining a Formula for an Inverse

If a function f is one-to-one, a formula for its inverse can generally be found as follows.

1. Replace $f(x)$ with y.

2. Interchange x and y.

3. Solve for y.

4. Replace y with $f^{-1}(x)$.

Example 6   Determine whether the function $f(x) = 2x - 3$ is one-to-one, and if it is, find a formula for $f^{-1}(x)$.

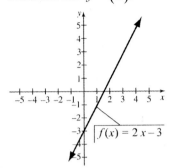

$f(x) = 2x - 3$

The graph passes the horizontal-line test, so it is one-to-one. Find the inverse.

1. Replace $f(x)$ with y:

   $\boxed{}$

2. Interchange x and y.

   $\boxed{}$

3. Solve for y:
   $$x + 3 = 2y$$

   $\boxed{} = y$

4. Replace y with $f^{-1}(x)$:

   $$f^{-1}(x) = \boxed{}$$

Example 7   Graph $f(x) = 2x - 3$ and $f^{-1}(x) = \dfrac{x+3}{2}$ using the same set of axes. Then compare the two graphs.

x	$f(x) = 2x - 3$
-1	-5
0	$\boxed{}$
2	$\boxed{}$
3	$\boxed{}$

x	$f^{-1}(x) = \dfrac{x+3}{2}$
-5	-1
-3	$\boxed{}$
1	$\boxed{}$
3	$\boxed{}$

The graphs are reflections of each other across the line $y = x$.

The graph of f^{-1} is a reflection of the graph of f across the line $y = x$.

Example 8   Consider $g(x) = x^3 + 2$.

a) Determine whether the function is one-to-one.

b) If it is one-to-one, find a formula for its inverse.

c) Graph the function and its inverse.

a) $g(x)$ passes the horizontal-line test, so it is one-to-one.

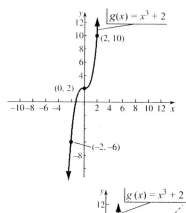

b) **1.** Replace $g(x)$ with y.

$$y = x^3 + 2$$

2. Interchange x and y.

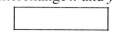

3. Solve for y.

$$x = y^3 + 2$$

$$\boxed{} = y^3$$

$$\boxed{} = y$$

c)

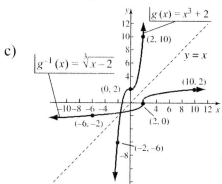

4. Replace y with $g^{-1}(x)$

$$g^{-1}(x) = \boxed{}$$

If a function f is one-to-one, then f^{-1} is the unique function such that each of the following holds:

$\left(f^{-1} \circ f\right)(x) = f^{-1}\left(f(x)\right) = x,$ for each x in the domain of f, and

$\left(f \circ f^{-1}\right)(x) = f\left(f^{-1}(x)\right) = x,$ for each x in the domain of f^{-1}.

Example 9   Given that $f(x) = 5x + 8$, use composition of functions to show that $f^{-1}(x) = \dfrac{x-8}{5}$.

$$\left(f^{-1} \circ f\right)(x) = f^{-1}\left(f(x)\right) = f^{-1}\left(\boxed{}\right) = \frac{\left(\boxed{} - 8\right)}{5} = \frac{5x}{5} = \boxed{}$$

$$\left(f \circ f^{-1}\right)(x) = f\left(f^{-1}(x)\right) = f\left(\boxed{}\right) = 5\left(\boxed{}\right) + 8 = x - 8 + 8 = \boxed{}$$

These two compositions show us that $f^{-1}(x) = \dfrac{x-8}{5}$ is the inverse of $f(x) = 5x + 8$.

Section 5.2   Exponential Functions and Graphs

Exponential Function

The function $f(x) = a^x$, where x is a real number, $a > 0$ and $a \neq 1$, is called the **exponential function, base a.**

Example 1   Graph the exponential function $y = f(x) = 2^x$.

x	$y = f(x) = 2^x$	(x, y)
0	1	$(0, 1)$
1	☐	$(1, 2)$
2	4	$\left(2, \boxed{}\right)$
3	8	$(3, 8)$
-1	$\dfrac{1}{2}$	$\left(-1, \dfrac{1}{2}\right)$
-2	☐	$\left(-2, \dfrac{1}{4}\right)$
-3	$\dfrac{1}{8}$	$\left(\boxed{}, \dfrac{1}{8}\right)$

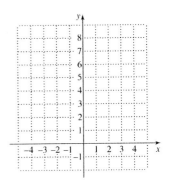

Example 2   Graph the exponential function $y = f(x) = \left(\dfrac{1}{2}\right)^x$.

Points of $g(x) = 2^x$	Points of $f(x) = \left(\dfrac{1}{2}\right)^x = 2^{-x}$
$(0, 1)$	$(0, 1)$
$(1, 2)$	$(-1, 2)$
$(2, 4)$	$\left(\boxed{}, 4\right)$
$(3, 8)$	$(-3, 8)$
$\left(-1, \dfrac{1}{2}\right)$	$\left(\boxed{}, \dfrac{1}{2}\right)$
$\left(-2, \dfrac{1}{4}\right)$	$\left(2, \dfrac{1}{4}\right)$
$\left(-3, \dfrac{1}{8}\right)$	$\left(3, \dfrac{1}{8}\right)$

$$y = f(x) = \left(\frac{1}{2}\right)^x = 2^{-x}$$
$$y = g(x) = 2^x$$

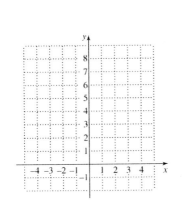

Example 3   Graph each of the following by hand. Then check your graph with a graphing calculator. Before doing so, describe how each graph can be obtained from the graph of $f(x) = 2^x$.

a) $f(x) = 2^{x-2}$ b) $f(x) = 2^x - 4$ c) $f(x) = 5 - 0.5^x$

a) Graph $f(x) = 2^{x-2}$.

The graph of $f(x) = 2^{x-2}$ is the graph of $f(x) = 2^x$ shifted _____ 2 units.
$$\underset{\text{left / right}}{}$$

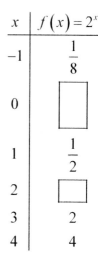

x	$f(x) = 2^{x-2}$
-1	$\dfrac{1}{8}$
0	☐
1	$\dfrac{1}{2}$
2	☐
3	2
4	4

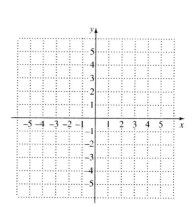

b) Graph $f(x) = 2^x - 4$.

Next, we consider the graph of $f(x) = 2^x - 4$, which is the graph of $f(x) = 2^x$ shifted _____ 4 units.
$$\underset{\text{up / down}}{}$$

x	$f(x) = 2^x - 4$
-2	$-3\dfrac{3}{4}$
-1	☐
0	-3
1	☐
2	0
3	4

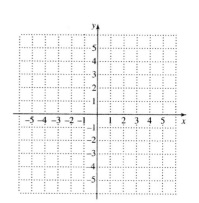

c) Graph $f(x) = 5 - 0.5^x$.

$$f(x) = 5 - 0.5^x = 5 - \left(\frac{1}{2}\right)^x$$

$$= 5 - \left(2^{-1}\right)^x$$

$$= 5 - 2^{-x}$$

The graph of $f(x) = 5 - 0.5^x$ is the graph of $f(x) = 2^x$ reflected across the

_____ , then reflected across the _____ , and finally shifted
   x-axis / y-axis   x-axis / y-axis

_____ , 5 units.
   up / down

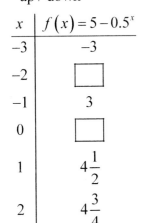

x	$f(x) = 5 - 0.5^x$
−3	−3
−2	☐
−1	3
0	☐
1	$4\frac{1}{2}$
2	$4\frac{3}{4}$

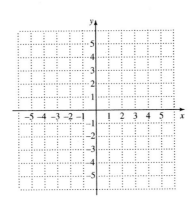

Example 4   *Compound Interest.* The amount of money A to which a principal P will grow after t years at interest rate r (in decimal form), compounded n times per year, is given by the formula

$$A = P\left(1 + \frac{r}{n}\right)^{nt}.$$

Suppose that $100,000 is invested at 6.5% interest, compounded semiannually.

a) Find a function for the amount to which the investment grows after t years.

$$A = P\left(1 + \frac{r}{n}\right)^{nt}$$

$$P = 100,000 \qquad r = \boxed{} \qquad n = 2$$

$$A(t) = 100,000\left(1 + \frac{0.065}{2}\right)^{2t}$$

$$A(t) = 100,000\left(\boxed{}\right)^{2t}$$

b) Graph the function.
We can graph the function on a graphing calculator or by hand.

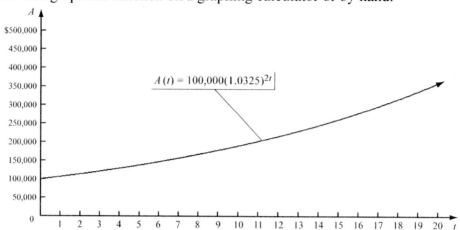

$A(t) = 100,000(1.0325)^{2t}$

c) Find the amount of money in the account at $t = 0, 4, 8$, and 10 years.

$A(0) = 100,000(1.0325)^{2 \cdot \square} = \boxed{}$

$A(4) = 100,000(1.0325)^{2 \cdot 4} \approx \$129,157.75$

$A(8) = 100,000(1.0325)^{2 \cdot \square} \approx \boxed{}$

$A(10) = 100,000(1.0325)^{2 \cdot 10} \approx \$189,583.79$

We could also find these values using function notation $Y1(0)$, $Y1(4)$, $Y1(8)$, and $Y1(10)$ on the calculator's home screen after entering the function as $Y1$.

d) When will the amount of money in the account reach $400,000?

We solve $400,000 = 100,000(1.0325)^{2t}$. On the calculator, graph $Y2 = 400,000$ along with $Y1 = 100,000(1.0325)^{2t}$. We will use the window $[0, \ 30, 0, \ 500,000]$, $\text{Xscl} = 5$, $\text{Yscl} = 50,000$. Then use the Intersect feature to find the coordinates of the point of intersection of the graphs. The first coordinate of this point, about $\boxed{}$, gives us the solution.

We can also use the Zero method to estimate the zero of $y = 100,000(1.0325)^{2t} - \boxed{}$. We change the viewing window to $[0, 30, -500,000, 500,000]$, $\text{Xscl} = 5$, $\text{Yscl} = 100,000$. Now use the Zero feature. The zero is approximately $\boxed{}$.

In either case we find that the account grows to $400,000 after about $\boxed{}$ years, or about 21 years and 8 months.

$e = 2.7182818284...$

Example 5   Find each value of e^x, to four decimal places, using the e^x key on a calculator.

a) $e^3 = e^{\wedge}(3) \approx$ ⬚

b) $e^{-0.23} = e^{\wedge}(-0.23) \approx$ ⬚

c) $e^0 = e^{\wedge}(0) =$ ⬚

Example 6   Graph $f(x) = e^x$ and $g(x) = e^{-x}$.

x	$f(x) = e^x$	$g(x) = e^{-x}$
-2	0.135	7.389
-1	0.368	⬚
0	⬚	1
1	2.718	0.368
2	7.389	0.135

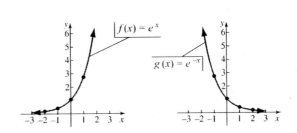

Example 7   Graph each of the following. Before doing so, describe how each graph can be obtained from the graph of $y = e^x$.

a) $f(x) = e^{x+3}$

b) $f(x) = e^{-0.5x}$

c) $f(x) = 1 - e^{-2x}$

a)   Graph $f(x) = e^{x+3}$.

The graph of $f(x) = e^{x+3}$ is the graph of $f(x) = e^x$ shifted _____ 3 units.
                                                              left / right

x	$f(x) = e^{x+3}$
-5	0.135
-4	0.368
-3	⬚
-2	⬚
-1	7.389
0	20.086
1	54.598

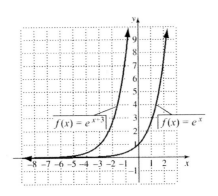

b) Graph $f(x) = e^{-0.5x}$.

The graph of $f(x) = e^{-0.5x}$ is the graph of $f(x) = e^x$ stretched

_____ and reflected across the _____.
horizontally / vertically x-axis / y-axis

x	$f(x) = e^{-0.5x}$
-3	4.482
-2	
-1	1.649
0	
1	0.607
2	0.368

c) Graph $f(x) = 1 - e^{-2x}$.

The graph of $f(x) = 1 - e^{-2x}$ is the graph of $f(x) = e^x$ shrunk

_____, reflected across the _____ and also reflected
horizontally / vertically x-axis / y-axis

across the _____, and finally shifted _____ 1 unit.
               x-axis / y-axis   up / down

x	$f(x) = 1 - e^{-2x}$
-1	-6.389
0	
1	0.865
2	0.982
3	

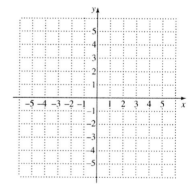

Section 5.3 Logarithmic Functions and Graphs

Example 1 Graph $x = 2^y$.

$x \ \left(x = 2^y\right)$	y	(x, y)
1	0	$(1, 0)$
☐	1	☐
4	2	$(4, 2)$
☐	-1	☐
☐	-2	☐
$\dfrac{1}{8}$	-3	$\left(\dfrac{1}{8}, -3\right)$

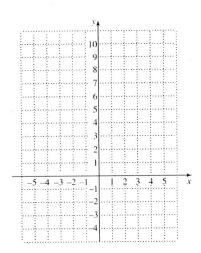

Logarithmic Function, Base a

We define $y = \log_a x$ as that number y such that $x = a^y$, where $x > 0$ and a is a positive constant other than 1.

Example 2

a) Find $\log_{10} 10{,}000$.

$$10^{\boxed{?}} = 10{,}000$$

$$\log_{10} 10{,}000 = \boxed{}$$

b) Find $\log_{10} 0.01$.

$$10^{\boxed{?}} = \frac{1}{100} = 10^{\boxed{}}$$

$$\log_{10} 0.01 = \boxed{}$$

c) Find $\log_5 125$.

$$5^{\boxed{?}} = 125$$

$$\log_5 125 = \boxed{}$$

d) Find $\log_9 3$.

$$9^{\boxed{?}} = 3 = 9^{1/2}$$

$$\log_9 3 = \boxed{}$$

e) Find $\log_6 1$.

$$6^{\boxed{?}} = 1$$

$$\log_6 1 = \boxed{}$$

f) Find $\log_5 5$.

$$5^{\boxed{?}} = 5$$

$$\log_5 5 = \boxed{}$$

$\log_a 1 = 0$ and $\log_a a = 1$, for any logarithmic base a.

$\log_a x = y \leftrightarrow x = a^y$ A logarithm is an exponent!

Example 3

Convert each of the following to a logarithmic equation.

a) $16 = 2^x$ $\log_{\square} 16 = \boxed{}$

b) $10^{-3} = 0.001$ $\log_{\square} 0.001 = \boxed{}$

c) $e^t = 70$ $\log_{\square} 70 = \boxed{}$

Example 4

Convert each of the following to an exponential equation.

a) $\log_2 32 = 5$ $2^{\square} = \boxed{}$

b) $\log_a Q = 8$ $a^{\square} = \boxed{}$

c) $x = \log_t M$ $t^{\square} = \boxed{}$

Common Logarithms (Base 10): $\log M = \log_{10} M$

Natural Logarithms (Base e): $\ln M = \log_e M$

Example 5   Find each of the following common logarithms on a calculator. If you are using a graphing calculator, set the calculator in REAL mode. Round to four decimal places.

a) $\log(645778) \approx \boxed{}$

b) $\log(.0000239) \approx \boxed{}$

c) $\log(-3)$

The domain of the log function is $(0, \infty)$; -3 is not in this domain.

$\log(-3)$ _____ exist.
                does / does not

Example 6 Find each of the following natural logarithms on a calculator. If you are using a graphing calculator, set the calculator in REAL mode. Round to four decimal places.

a) $\ln(645,778) \approx$ ☐

b) $\ln(0.000039) \approx$ ☐

c) $\ln(-5)$

The domain of the ln function is $(0, \infty)$; -5 is not in this domain.

$\ln(-5)$ _____ exist.
　　　　　 does / does not

d) $\ln(e) = 1$ 　　　　　　　　　$e^? = e$

　　　　　　　　　　　　　　　　$e^{\square} = e$

e) $\ln(1) = 0$ 　　　　　　　　　$e^? = 1$

　　　　　　　　　　　　　　　　$e^{\square} = 1$

$\ln 1 = 0$ and $\ln e = 1$, for the logarithmic base e.

The Change-of-Base Formula

For any logarithmic bases a and b, and any positive number M,

$$\log_b M = \frac{\log_a M}{\log_a b}.$$

Example 7 Find $\log_5 8$ using common logarithms.

Change-of-base formula:

$$\log_b M = \frac{\log_a M}{\log_a b}$$

Let $a = $ ☐ , $b = $ ☐ , and $M = 8$.

$$\log_5 8 = \frac{\log_{10} \boxed{}}{\log_{10} 5} \approx \boxed{}$$

Example 8   Find $\log_5 8$ using natural logarithms.

Change-of-base formula:

$$\log_b M = \frac{\log_a M}{\log_a b}$$

Let $a = \boxed{}$, $b = 5$, and $M = \boxed{}$.

$$\log_5 8 = \frac{\log_\square 8}{\log_\square 5} = \frac{\ln 8}{\ln 5} \approx \boxed{}$$

Example 9   Graph $y = f(x) = \log_5 x$.

Think: $x = \boxed{}$

x, or 5^y	y
1	0
$\boxed{}$	1
$\boxed{}$	−1
$\dfrac{1}{25}$	−2

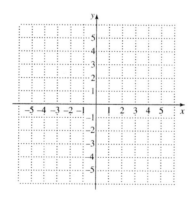

Example 10   Graph $g(x) = \ln x$.

x	$g(x) = \ln x$
0.5	−0.7
1	$\boxed{}$
2	$\boxed{}$
3	$\boxed{}$
4	1.4
5	1.6

Example 11   Graph each of the following. Before doing so, describe how each graph can be obtained from the graph of $y = \ln x$. Give the domain and the vertical asymptote of each function. Then check your graph with a graphing calculator.

a) $f(x) = \ln(x+3)$ b) $f(x) = 3 - \dfrac{1}{2}\ln x$ c) $f(x) = \left|\ln(x-1)\right|$

a) Graph $f(x) = \ln(x+3)$.

The graph of $f(x) = \ln(x+3)$ is a shift of the graph of $y = \ln x$ _____ 3 units.
<u>left / right</u>

x	$f(x)$
−2.9	−2.303
−2	0
0	1.099
2	1.609
4	1.946

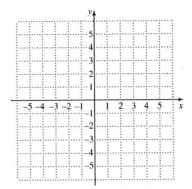

Domain: $\left(\boxed{}, \infty\right)$

Vertical asymptote: $\boxed{}$

b) Graph $f(x) = 3 - \dfrac{1}{2}\ln x$.

The graph of $f(x) = 3 - \dfrac{1}{2}\ln x$ is a _____ shrinking of the graph of
<u>vertical / horizontal</u>

$y = \ln x$, followed by a reflection across the _____ , and then a translation
<u>x-axis / y-axis</u>

_____ 3 units.
<u>up / down</u>

x	$f(x)$
0.1	4.151
1	3
3	2.451
6	2.104
9	1.901

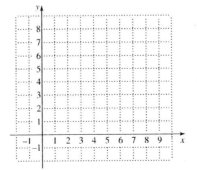

Domain: $\left(\boxed{}, \infty\right)$

Vertical asymptote: $\boxed{}$

c) Graph $f(x) = \left| \ln(x-1) \right|$.

The graph of $f(x) = \left| \ln(x-1) \right|$ is a shift of the graph of $y = \ln x$ _____ 1 unit
                                                                          left / right

followed by a reflection of negative outputs across the _____ .
                                                          x-axis / y-axis

x	$f(x)$
1.1	2.303
2	0
4	1.099
6	1.609
8	1.946

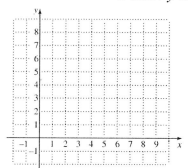

Domain: $\left(\boxed{}, \infty \right)$

Vertical asymptote: $\boxed{}$

Example 12   In a study by psychologists Bornstein and Bornstein, it was found that the average walking speed w, in feet per second, of a person living in a city of population P, in thousands, is given by the function

$$w(P) = 0.37 \ln P + 0.05.$$

a) The population of Billings, Montana, is 106,954. Find the average walking speed of people living in Billings.

$106{,}954 = 106.954$ thousands

$w(106.954) = 0.37 \ln \left(\boxed{} \right) + 0.05$

$\approx \boxed{}$ ft/sec

b) The population of Chicago, Illinois, is 2,714,856. Find the average walking speed of people living in Chicago.

$2{,}714{,}856 = 2714.856$ thousands

$w(2714.856) = 0.37 \ln \left(\boxed{} \right) + 0.05$

$\approx \boxed{}$ ft/sec

c) Graph the function.

On a graphing calculator, enter $Y1 = 0.37 \ln(X) + 0.05$, and choose an appropriate window. We will use $[0, 600, 0, 4]$, $Xscl = 100$. Press GRAPH to see the graph. (Sketch the graph in the space below.)

d) A sociologist computes the average walking speed in a city to be approximately 2.0 ft/sec. Use this information to estimate the population of the city.

We will solve $2.0 = 0.37 \ln P + 0.05$ using the INTERSECT method on a graphing calculator. We graph $Y1 = 0.37 \ln(X) + 0.05$ and $Y2 = 2.0$ and find the point of intersection of the graphs. The first coordinate of the point of intersection is approximately 194.5. This tells us that in a city in which the average walking speed is 2.0 ft/sec, the population is about 194.5 thousand, or 194,500.

Example 13   Measured on the Richter scale, the magnitude R of an earthquake of intensity I is defined as

$$R = \log \frac{I}{I_0},$$

where I_0 is a minimum intensity used for comparison. We can think of I_0 as a threshold intensity that is the weakest earthquake that can be recorded on a seismograph. If one earthquake is 10 times as intense as another, its magnitude on the Richter scale is 1 greater than that of the other. If one earthquake is 100 times as intense as another, its magnitude on the Richter scale is 2 higher, and so on. Thus an earthquake whose magnitude is 5 on the Richter scale is 10 times as intense as an earthquake whose magnitude is 4. Earthquake intensities can be interpreted as multiples of the minimum intensity I_0.

The Napal region earthquake in South Asia on April 25, 2015, had an intensity of $10^{7.8} \cdot I_0$. It caused extensive loss of life and severe structural damage to buildings, railways, and roads. What was the magnitude of this earthquake on the Richter scale?

$$R = \log \frac{I}{I_0} \qquad\qquad I = 10^{7.8} \cdot I_0$$

$$R = \log \frac{\boxed{}}{I_0}$$

$$R = \log \boxed{}$$

$$R = 7.8$$

The magnitude of the earthquake was $\boxed{}$ on the Richter scale.

Section 5.4   Properties of Logarithmic Functions

Properties of Logarithms

For any positive numbers M and N, any logarithmic base a, and any real number p:

    Product Rule: $\log_a MN = \log_a M + \log_a N.$

    Power Rule: $\log_a M^p = p \log_a M.$

    Quotient Rule: $\log_a \dfrac{M}{N} = \log_a M - \log_a N.$

Example 1   Express as a sum of logarithms.

$\log_3 (9 \cdot 27) = \boxed{} + \boxed{}$ Using the product rule

Example 2   Express as a single logarithm.

$\log_2 p^3 + \log_2 q = \log_2 \left(\boxed{} \right)$ Using the product rule

Example 3

a) Express $\log_a 11^{-3}$ as a product.

    Power Rule: $\log_a M^p = p \log_a M$

    $M = 11$ and $p = -3$

    $\log_a 11^{-3} = \boxed{} \log_a \boxed{}$

b) Express $\log_a \sqrt[4]{7}$ as a product.

    First rewrite $\sqrt[4]{7}$ with an exponent using $\sqrt[n]{x} = x^{\frac{1}{n}}.$

    $\log_a \sqrt[4]{7} = \log_a 7^{\boxed{}}$

    $= \boxed{} \log_a 7$ Using the power rule

c) Express $\ln x^6$ as a product.

    Using the power rule:

    $\ln x^6 = \boxed{} \ln \boxed{}$

Example 4

Express as a difference of logarithms.

$$\log_t \frac{8}{w} = \log_t \boxed{} - \log_t \boxed{} \qquad \text{Using the quotient rule}$$

Example 5

Express as a single logarithm.

$$\log_b 64 - \log_b 16 = \boxed{} \qquad \text{Using the quotient rule}$$

$$= \boxed{}$$

Example 6

a) Express $\log_a \dfrac{x^2 y^5}{z^4}$ in terms of sums and differences of logarithms.

$$\log_a \frac{x^2 y^5}{z^4} = \log_a \left(\boxed{} \right) - \log_a \boxed{}$$

$$= \log_{\boxed{}} x^2 + \log_{\boxed{}} y^5 - \log_a z^4$$

$$= \boxed{}\log_a x + \boxed{}\log_a y - \boxed{}\log_a z$$

b) Express $\log_a \sqrt[3]{\dfrac{a^2 b}{c^5}}$ in terms of sums and differences of logarithms.

$$\log_a \sqrt[3]{\frac{a^2 b}{c^5}} = \log_a \left(\frac{a^2 b}{c^5} \right)^{\boxed{}}$$

$$= \boxed{} \log_a \left(\frac{a^2 b}{c^5} \right)$$

$$= \frac{1}{3} \left(\log_a \left(\boxed{} \right) - \log_a \boxed{} \right)$$

$$= \frac{1}{3} \left(\log_a \boxed{} + \log_a \boxed{} - \log_a c^5 \right)$$

$$= \frac{1}{3} \left(\boxed{} \log_a a + \log_a b - \boxed{} \log_a c \right)$$

$$= \boxed{} + \boxed{} \log_a b - \boxed{} \log_a c$$

c) Express $\log_b \dfrac{ay^5}{m^3 n^4}$ in terms of sums and differences of logarithms.

$$\log_b \dfrac{ay^5}{m^3 n^4} = \log_b \left(ay^5\right) - \log_{\square} \left(m^3 n^4\right)$$

$$= \log_b \boxed{} + \log_b \boxed{} - \left(\log_{\square} m^3 + \log_{\square} n^4\right)$$

$$= \log_b a + \log_b y^5 \boxed{} \log_b m^3 \boxed{} \log_b n^4$$

$$= \log_b a + \boxed{} \log_b y - \boxed{} \log_b m - \boxed{} \log_b n$$

Example 7   Express as a single logarithm.

$$5\log_b x - \log_b y + \frac{1}{4}\log_b z$$

$$= \log_b x^{\boxed{}} - \log_b y + \log_b z^{\boxed{}}$$

$$= \log_b \dfrac{\boxed{}}{\boxed{}} + \log_b z^{\frac{1}{4}}$$

$$= \log_b \dfrac{\boxed{}}{\boxed{}}$$

Example 8   Express as a single logarithm.

$$\ln\left(3x+1\right) - \ln\left(3x^2 - 5x - 2\right)$$

$$= \ln \dfrac{\boxed{}}{3x^2 - 5x - 2} = \ln \dfrac{3x+1}{(3x+1)\left(\boxed{}\right)}$$

$$= \ln \dfrac{1}{\boxed{}}$$

Example 9

a) Given that $\log_a 2 \approx 0.301$ and $\log_a 3 \approx 0.477$, find $\log_a 6$, if possible.

$$\log_a 6 = \log_a \left(2 \cdot 3\right)$$

$$= \log_a \boxed{} + \log_a \boxed{}$$

$$\approx 0.301 + \boxed{}$$

$$\approx \boxed{}$$

b) Given that $\log_a 2 \approx 0.301$ and $\log_a 3 \approx 0.477$, find $\log_a \dfrac{2}{3}$, if possible.

$$\log_a \frac{2}{3} = \log_a 2 - \log_a \boxed{}$$

$$\approx \boxed{} - 0.477$$

$$\approx \boxed{}$$

c) Given that $\log_a 2 \approx 0.301$ and $\log_a 3 \approx 0.477$, find $\log_a 81$, if possible.

$$\log_a 81 = \log_a \boxed{}$$

$$= \boxed{} \log_a 3$$

$$\approx 4 \left(\boxed{} \right)$$

$$\approx 1.908$$

d) Given that $\log_a 2 \approx 0.301$ and $\log_a 3 \approx 0.477$, find $\log_a \dfrac{1}{4}$, if possible.

$$\log_a \frac{1}{4} = \log_a 1 - \log_a \boxed{}$$

$$= \boxed{} - \log_a 2^2$$

$$= -2 \log_a \boxed{}$$

$$\approx -2 \left(\boxed{} \right)$$

$$\approx -0.602$$

e) Given that $\log_a 2 \approx 0.301$ and $\log_a 3 \approx 0.477$, find $\log_a 5$, if possible.

$$\log_a 5 = \log_a (2 + 3) \neq \log_a 2 + \log_a 3$$

Finding $\log_a 5$ using the given logarithms _____ possible.
$$\text{is / is not}$$

f) Given that $\log_a 2 \approx 0.301$ and $\log_a 3 \approx 0.477$, find $\dfrac{\log_a 3}{\log_a 2}$, if possible.

$$\frac{\log_a 3}{\log_a 2} \approx \frac{\boxed{}}{\boxed{}} \approx 1.585$$

The Logarithm of a Base to a Power

For any base a and any positive real number x,

$\log_a a^x = x.$

Example 10

a) Simplify.

$\log_a a^8 = \boxed{}$

b) Simplify.

$\ln e^{-t} = \log_{\boxed{}} e^{-t}$

$= \boxed{}$

c) Simplify.

$\log 10^{3k} = \log_{\boxed{}} 10^{3k}$

$= \boxed{}$

A Base to a Logarithmic Power

For any base a and any positive real number x,

$a^{\log_a x} = x.$

Example 11

a) Simplify.

$4^{\log_4 k} = \boxed{}$

b) Simplify.

$e^{\ln 5} = e^{\boxed{}} = \boxed{}$

c) Simplify.

$10^{\log 7t} = 10^{\boxed{}} = \boxed{}$

Section 5.5 Solving Exponential Equations and Logarithmic Equations

Base-Exponent Property

For any $a > 0$, $a \neq 1$,

$$a^x = a^y \leftrightarrow x = y.$$

Example 1   Solve $2^{3x-7} = 32$.

$2^{3x-7} = 32$

$2^{3x-7} = \boxed{}$

$3x - 7 = 5$

$3x = \boxed{}$

$x = \boxed{}$

Check:

$$\frac{2^{3x-7} = 32}{2^{3\boxed{}-7} \overset{?}{=} 32}$$

$2^{\boxed{}}$

$32 \mid 32$ TRUE

The solution is $\boxed{}$.

Property of Logarithmic Equality

For any $M > 0$, $N > 0$, $a > 0$, and $a \neq 1$,

$$\log_a M = \log_a N \leftrightarrow M = N.$$

Example 2   Solve $3^x = 20$.

$3^x = 20$

$\log 3^x = \log 20$

$\boxed{} \log 3 = \log 20$

$x = \dfrac{\log 20}{\boxed{}}$

$x \approx 2.7268$

Check: $3^{2.7268} \approx \boxed{}$

The solution is about $\boxed{}$.

Example 3   Solve $100e^{0.08t} = 2500$.

$$100e^{0.08t} = 2500$$

$$e^{0.08t} = \boxed{}$$

$$\ln e^{0.08t} = \ln 25$$

$$\boxed{} = \ln 25$$

$$t = \frac{\ln 25}{\boxed{}}$$

$$t \approx 40.2$$

The solution is about $\boxed{}$.

Example 4   Solve $4^{x+3} = 3^{-x}$.

$$4^{x+3} = 3^{-x}$$

$$\log 4^{x+3} = \log 3^{-x}$$

$$\left(\boxed{}\right)\log 4 = \boxed{}\log 3$$

$$x\log 4 + \boxed{}\log 4 = -x\log 3$$

$$x\log 4 + \boxed{} = -3\log 4$$

$$\boxed{}(\log 4 + \log 3) = -3\log 4$$

$$x = \frac{-3\log 4}{\boxed{}}$$

$$x \approx -1.6737$$

The solution is about $\boxed{}$.

Example 5 Solve $e^x + e^{-x} - 6 = 0$.

$$e^x + e^{-x} - 6 = 0$$

$$e^x + \frac{1}{\boxed{}} - 6 = 0$$

$$\boxed{} \left(e^x + \frac{1}{e^x} - 6 \right) = \boxed{} \cdot 0$$

$$e^{2x} + 1 - 6e^x = 0$$

$$e^{2x} - 6e^x + 1 = 0 \qquad\qquad e^{2x} = \left(e^x \right)^2$$

$$\left(\boxed{} \right)^2 - 6e^x + 1 = 0 \qquad\qquad \text{Let } u = e^x.$$

$$u^2 - 6 \cdot \boxed{} + 1 = 0$$

$$a = 1 \qquad b = -6 \qquad c = 1$$

$$u = \frac{-b \pm \sqrt{b^2 - 4ac}}{2a}$$

$$= \frac{-\left(\boxed{} \right) \pm \sqrt{(-6)^2 - 4 \cdot 1 \cdot 1}}{2 \cdot 1}$$

$$= \frac{6 \pm \sqrt{36 - 4}}{2}$$

$$= \frac{6 \pm \sqrt{\boxed{}}}{2} = \frac{6 \pm \boxed{} \cdot \sqrt{2}}{2} = 3 \pm \boxed{}$$

$$e^x = 3 \pm 2\sqrt{2}$$

$$\ln e^x = \boxed{} \left(3 \pm 2\sqrt{2} \right)$$

$$\boxed{} \cdot \ln e = \ln \left(3 \pm 2\sqrt{2} \right)$$

$$x = \ln \left(3 \pm 2\sqrt{2} \right)$$

The solutions are approximately $\boxed{}$ and $\boxed{}$.

Example 6   Solve $\log_3 x = -2$.

$\log_3 x = -2$

$3^{\boxed{}} = x$

$\dfrac{1}{3^2} = x$

$\boxed{} = x$

Check:

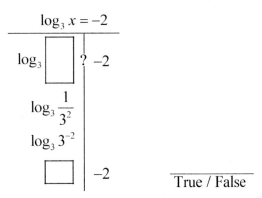

The solution is $\dfrac{1}{9}$.

Example 7   Solve $\log x + \log(x + 3) = 1$.

$\log x + \log(x + 3) = 1$

$\log_{10}\left[x\left(\boxed{}\right)\right] = 1$

$x(x + 3) = \boxed{}^1$

$x^2 + \boxed{} = 10$

$x^2 + 3x - 10 = 0$

$(x - 2)\left(\boxed{}\right) = 0$

$x - 2 = 0 \quad$ or $\quad x + 5 = 0$

$x = \boxed{} \quad$ or $\quad x = \boxed{}$

Check: For 2:

$\log x + \log(x + 3) = 1$

$\log\boxed{} + \log\left(\boxed{} + 3\right) \;?\; 1$

$\log 2 + \log 5$

$\log(2 \cdot 5)$

$\log 10$

$1 \;\big|\; 1 \quad$ TRUE

Check: For -5:

$\log x + \log(x + 3) = 1$

$\log\left(\boxed{}\right) + \log(-5 + 3) \;?\; 1$

We see that -5 _____ be a
               can / cannot

solution because the logarithmic
function is not defined for negative
values of x.

The solution is $\boxed{}$.

Example 8 Solve $\log_3(2x-1) - \log_3(x-4) = 2$.

$$\log_3(2x-1) - \log_3(x-4) = 2$$

$$\log_3 \frac{2x-1}{\boxed{}} = 2$$

$$\frac{2x-1}{x-4} = \boxed{}^2$$

$$\frac{2x-1}{x-4} = 9$$

$$(x-4) \cdot \frac{2x-1}{x-4} = (x-4) \cdot 9$$

$$2x-1 = 9x-36$$

$$35 = 7x$$

$$5 = x$$

Check:

$$\begin{array}{c} \log_3(2x-1) - \log_3(x-4) = 2 \\ \hline \log_3(2\cdot5-1) - \log_3(5-4) \overset{?}{=} 2 \\ \log_3\boxed{} - \log_3\boxed{} \\ 2 - \boxed{} \\ 2 \,\big|\, 2 \quad \text{TRUE} \end{array}$$

The solution is $\boxed{}$.

Example 9 Solve $\ln(4x+6) - \ln(x+5) = \ln x$.

$$\ln(4x+6) - \ln(x+5) = \ln x$$

$$\ln \frac{\boxed{}}{x+5} = \ln x$$

$$\frac{4x+6}{x+5} = \boxed{}$$

$$(x+5) \cdot \frac{4x+6}{x+5} = (x+5) \cdot x$$

$$4x+6 = \boxed{} + 5x$$

$$0 = x^2 + x - 6$$

$$0 = (x+3)\left(\boxed{}\right)$$

$$x+3 = 0 \qquad or \quad x-2 = 0$$

$$x = \boxed{} \quad or \qquad x = \boxed{}$$

The number -3 $\underset{\text{is / is not}}{\underline{}}$ a solution because $4(-3)+6 = -6$ and $\ln(-6)$ is not a real

number. The value $\boxed{}$ checks and is the solution.

Example 10 Solve: $e^{0.5x} - 7.3 = 2.08x + 6.2$.

Using the Intersect Method:

Enter $Y1 = e^{0.5x} - 7.3$ and $Y2 = 2.08x + 6.2$. When we graph these equations, we see $\boxed{}$ points of intersection. Use the Intersect feature twice. The $\underset{\text{first / second}}{\underline{}}$ coordinates of the points of intersection are the solutions of the equation. They are approximately -6.471 and $\boxed{}$.

Using the Zero method:

First rewrite the equation with 0 on one side. We have $e^{0.5x} - 7.3 - 2.08x - 6.2 = 0$. On the calculator, enter $Y1 = e^{0.5x} - 7.3 - 2.08x - 6.2$. Graph the equation and use the Zero feature to find the zeros. They are approximately $\boxed{}$ and 6.610.

These are the solutions of the equation.

Section 5.6   Applications and Models: Growth and Decay; Compound Interest

The **population growth function** is

$$P(t) = P_0 e^{kt}, \ k > 0,$$

where P_0 is the population at time 0, $P(t)$ is the population after time t, and k is the **exponential growth rate**.

Example 1   In 2014, the population of Ghana, located on the west coast of Africa, was about 25.8 million, and the exponential growth rate was 2.19% per year.

a)   Find the exponential growth function.

$P(t) = 25.8e^{\boxed{} \cdot t}$, where t = number of years after 2014 and $P(t)$ is in millions

b)   Graph the exponential growth function.

Graph $Y1 = 25.8e^{0.219x}$ in the window $[0, 80, 0, 200]$, Xscl $= 10$, Yscl $= 25$. (Sketch the graph in the space below.)

c)   Estimate the population in 2018.

$P\left(\boxed{}\right) = 25.8e^{0.0219\left(\boxed{}\right)} \approx 28.2$ million

We can also use the Value feature on a graphing calculator to find $P(4)$.

d) At this growth rate, when will the population be 40 million?

$$\boxed{} = 25.8e^{0.0219t}$$

$$\frac{40}{25.8} = e^{0.0219t}$$

$$\ln\left(\frac{40}{25.8}\right) = \boxed{}$$

$$\frac{\ln\left(\dfrac{40}{25.8}\right)}{\boxed{}} = t$$

$$20 \approx t$$

The population of Ghana will be 40 million about $\boxed{}$ years after 2014.

Interest Compounded Continuously

When an amount P_0 is invested in a savings account at interest rate r compounded continuously, the amount $P(t)$ in the account after t years is given by the function

$$P(t) = P_0 e^{kt}, \ k > 0.$$

Example 2 Suppose that $2000 is invested at interest rate k, compounded continuously, and grows to $2504.65 in 5 years.

a) What is the interest rate?

$$P(t) = 2000e^{kt}$$

$$\boxed{} = 2000e^{k\cdot\boxed{}}$$

$$\frac{2504.65}{2000} = e^{5k}$$

$$\ln\frac{2504.65}{2000} = \ln e^{5k}$$

$$\ln\frac{2504.65}{2000} = \boxed{}$$

$$\frac{\ln\dfrac{2504.65}{2000}}{\boxed{}} = k$$

$$0.045 \approx k$$

The interest rate is about $\boxed{}$.

b) Find the exponential growth function.

$$P(t) = P_0 e^{kt}$$

$$P(t) = \boxed{} e^{\boxed{} \cdot t}$$

c) What will the balance be after 10 years?

$$P(t) = 2000 e^{0.045t}$$

$$P(10) = 2000 e^{0.045(\boxed{})}$$

$$= 2000 e^{\boxed{}}$$

$$\approx 3136.62$$

The balance after 10 years is $\boxed{}$.

d) After how long will the $2000 have doubled?

$$\boxed{} = 2000 e^{0.045T}$$

$$\boxed{} = e^{0.045T}$$

$$\ln 2 = \ln e^{0.045T}$$

$$\ln 2 = \boxed{}$$

$$\frac{\ln 2}{\boxed{}} = T$$

$$15.4 \approx T$$

The investment of $2000 will double in about $\boxed{}$ years.

Growth Rate and Doubling Time

The **growth rate k** and the **doubling time T** are related by

$$kT = \ln 2, \text{ or } k = \frac{\ln 2}{T}, \text{ or } T = \frac{\ln 2}{k}.$$

Example 3   The population of the Philippines is now doubling every 37.7 years. What is the exponential growth rate?

$$k = \frac{\ln 2}{T}$$

$$k = \frac{\ln 2}{\boxed{}} \approx \boxed{} \approx 1.84\%$$

The growth rate of the population of the Philippines is about $\boxed{}$ per year.

Example 4   A lake is stocked with 400 fish of a new variety. The size of the lake, the availability of food, and the number of other fish restrict the growth of that type of fish in the lake to a limiting value of 2500. The population gets closer and closer to this limiting value, but never reaches it. The population of fish in the lake after time t, in months, is given by the function

$$P(t) = \frac{2500}{1 + 5.25e^{-0.32t}}.$$

a) Graph the function.

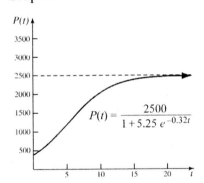

$$P(t) = \frac{2500}{1 + 5.25\,e^{-0.32t}}$$

Note that this function increases toward a limiting value of 2500. The graph has $y = 2500$ as a horizontal asymptote.

b) Find the population after 0, 1, 5, 10, 15, and 20 months.

These values can be found on a graphing calculator using a table set in ASK mode.

$P(0) = \boxed{}$

$P(1) \approx \boxed{}$

$P(5) \approx \boxed{}$

$P(10) \approx \boxed{}$

$P(15) \approx \boxed{}$

$P(20) \approx \boxed{}$

The **exponential decay function** is

$$P(t) = P_0 e^{-kt}, \ k > 0,$$

where P_0 is the population at time t, and k is the **decay rate**.

Decay Rate and Half-Life

The decay rate k and the half-life T are related by

$$kT = \ln 2, \ \text{or} \ k = \frac{\ln 2}{T}, \ \text{or} \ T = \frac{\ln 2}{k}.$$

Example 5   The radioactive element carbon-14 has a half-life of 5750 years. The percentage of carbon-14 present in the remains of organic matter can be used to determine the age of that organic matter. Archaeologists discovered that the linen wrapping from one of the Dead Sea Scrolls had lost 22.3% of its carbon-14 at the time it was found. How old was the linen wrapping?

$$P(t) = P_0 e^{-kt}, \ k > 0$$

Find k:

$$k = \frac{\ln 2}{T} = \frac{\ln 2}{\boxed{}} \approx 0.00012$$

$$P(t) = P_0 e^{\boxed{} \cdot t}$$

$$\boxed{} = P_0 e^{-0.00012t} \qquad\qquad \text{(77.7\% of the carbon-14 remains.)}$$

$$0.777 P_0 = P_0 e^{-0.00012t}$$

$$0.777 = \boxed{}$$

$$\ln 0.777 = \ln e^{-0.00012t}$$

$$\ln 0.777 = \boxed{}$$

$$\frac{\ln 0.777}{\boxed{}} = t$$

$$2103 \approx t$$

The linen wrapping on the Dead Sea Scrolls was about $\boxed{}$ years old.

Section 6.1   Trigonometric Functions of Acute Angles

Example 1   In the right triangle shown, find the six trigonometric function values of (a) θ and (b) α.

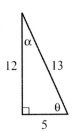

a) $\sin\theta = \dfrac{\text{opp}}{\text{hyp}} = \dfrac{12}{\square}$, $\csc\theta = \dfrac{\text{hyp}}{\text{opp}} = \dfrac{13}{12}$,

$\cos\theta = \dfrac{\text{adj}}{\text{hyp}} = \dfrac{5}{13}$, $\sec\theta = \dfrac{\text{hyp}}{\text{adj}} = \dfrac{13}{\square}$,

$\tan\theta = \dfrac{\text{opp}}{\text{adj}} = \dfrac{\square}{5}$, $\cot\theta = \dfrac{\text{adj}}{\text{opp}} = \dfrac{\square}{12}$

b) $\sin\alpha = \dfrac{\text{opp}}{\text{hyp}} = \dfrac{\square}{13}$, $\csc\alpha = \dfrac{\text{hyp}}{\text{opp}} = \dfrac{13}{\square}$,

$\cos\alpha = \dfrac{\text{adj}}{\text{hyp}} = \dfrac{12}{13}$, $\sec\alpha = \dfrac{\text{hyp}}{\text{adj}} = \dfrac{13}{12}$,

$\tan\alpha = \dfrac{\text{opp}}{\text{adj}} = \dfrac{5}{\square}$, $\cot\alpha = \dfrac{\text{adj}}{\text{opp}} = \dfrac{\square}{5}$

If we know the values of the sine, cosine, and tangent functions of an angle, we can use these reciprocal relationships to find the values of the cosecant, secant, and cotangent functions of that angle.

Reciprocal Functions

$$\csc\theta = \dfrac{1}{\sin\theta}, \qquad \sec\theta = \dfrac{1}{\cos\theta}, \qquad \cot\theta = \dfrac{1}{\tan\theta}$$

Example 2   Given that $\sin\phi = \dfrac{4}{5}$, $\cos\phi = \dfrac{3}{5}$, and $\tan\phi = \dfrac{4}{3}$, find $\csc\phi$, $\sec\phi$, and $\cot\phi$.

$$\csc\phi = \dfrac{1}{\sin\phi} = \dfrac{1}{\frac{4}{5}} = \dfrac{5}{\square}, \qquad \sec\phi = \dfrac{1}{\cos\phi} = \dfrac{1}{\frac{3}{5}} = \dfrac{5}{3}, \qquad \cot\phi = \dfrac{1}{\tan\phi} = \dfrac{1}{\frac{4}{3}} = \dfrac{\square}{4}$$

Example 3   Given that $\sin \beta = \dfrac{\sqrt{21}}{5}$, $\cos \beta = \dfrac{2}{5}$, and $\tan \beta = \dfrac{\sqrt{21}}{2}$, find $\csc \beta$, $\sec \beta$, and $\cot \beta$.

$$\csc \beta = \dfrac{1}{\sin \beta} = \dfrac{5}{\sqrt{21}} = \dfrac{5}{\sqrt{21}} \cdot \dfrac{\sqrt{21}}{\sqrt{21}} = \dfrac{5\sqrt{21}}{\boxed{}}, \quad \sec \beta = \dfrac{1}{\cos \beta} = \dfrac{5}{\boxed{}},$$

$$\cot \beta = \dfrac{1}{\boxed{}} = \dfrac{2}{\sqrt{21}} = \dfrac{2}{\sqrt{21}} \cdot \dfrac{\boxed{}}{\sqrt{21}} = \dfrac{2\sqrt{21}}{21}$$

Example 4   If $\sin \beta = \dfrac{6}{7}$ and β is an acute angle, find the other five trigonometric function values of β.

We sketch a right triangle in which the hypotenuse has length 7 and the side opposite β has length 6 (since $\sin \beta = 6/7 = \text{opp}/\text{hyp}$). We want to find a.

$$a^2 + b^2 = c^2$$
$$a^2 + 6^2 = 7^2$$
$$a^2 + 36 = \boxed{}$$
$$a^2 = 13$$
$$a = \sqrt{\boxed{}}$$

Now we can label the triangle as shown below.

$$\sin \beta = \dfrac{6}{7}, \qquad\qquad \csc \beta = \dfrac{\boxed{}}{6},$$

$$\cos \beta = \dfrac{\sqrt{13}}{7}, \qquad\qquad \sec \beta = \dfrac{\boxed{}}{\sqrt{13}}, \text{ or } \dfrac{7\sqrt{13}}{\boxed{}},$$

$$\tan \beta = \dfrac{6}{\sqrt{13}}, \text{ or } \dfrac{\boxed{}\sqrt{13}}{13}, \qquad \cot \beta = \dfrac{\sqrt{13}}{\boxed{}}$$

We will often use the function values of 30°, 45°, and 60°.

	30°	45°	60°
sin	$1/2$	$\sqrt{2}/2$	$\sqrt{3}/2$
cos	$\sqrt{3}/2$	$\sqrt{2}/2$	$1/2$
tan	$\sqrt{3}/3$	1	$\sqrt{3}$

Example 5   Trajectories for fireworks involve variables such as the launch angle, the launch velocity, and the size of the shell. Using physics to calculate the trajectory, fireworks technicians know exactly the path of the shell. Launch angles vary from 45° to 90°. For every inch of the diameter of the shell, a firework can travel about 100 ft vertically and 70 ft horizontally. These distances can vary depending on air resistance. A 6-in. shell launched at an angle of 60° travels a horizontal distance of 390 ft. Approximate the height of the fireworks display. Round the answer to the nearest foot.

We make a sketch.

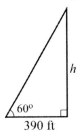

We have an angle and the side adjacent to it. We want to find the side h opposite it. We will use the tangent function to find h.

$$\tan 60° = \frac{h}{\boxed{}}$$

$$390 \cdot \tan 60° = h$$

$$390 \cdot \sqrt{3} = h$$

$$\boxed{} \approx h$$

The fireworks display is approximately 675 ft high.

Example 6   Convert $5°42'30''$ to decimal degree notation.

$$5°42'30'' = 5° + 42' + 30''$$

$$= 5° + 42' + \frac{30'}{\boxed{}} \qquad 1'' = \frac{1}{60}{}'$$

$$= 5° + 42' + 0.5'$$

$$= 5° + \boxed{}{}'$$

$$= 5° + \frac{42.5°}{60} \qquad 1' = \frac{1}{60}{}°$$

$$\approx 5.71° \qquad\qquad \frac{42.5°}{60} \approx 0.71°$$

Example 7   Convert 72.18° to D°M'S" notation.

$$72.18° = 72° + 0.18 \times 1°$$

$$= 72° + \boxed{} \times 60' \qquad 1° = 60'$$

$$= 72° + 10.8'$$

$$= 72° + 10' + \boxed{} \times 1'$$

$$= 72° + 10' + 0.8 \times 60" \qquad 1' = 60"$$

$$= 72° + 10' + \boxed{}"$$

$$= 72°10'48"$$

Example 8   Using a calculator, find the trigonometric function value, rounded to four decimal places, of each of the following.

a) $\tan 29.7°$ b) $\sec 48°$ c) $\sin 84°10'39"$

Be sure that the calculator is in DEGREE mode.

a) $\tan 29.7° \approx 0.5703899297 \approx 0.5704$

b) $\sec 48° = \dfrac{1}{\boxed{}\,48°} \approx 1.49447655 \approx 1.4945$

c) $\sin 84°10'39" \approx 0.9948409474 \approx 0.9948$

Example 9   Find the acute angle, to the nearest tenth of a degree, whose sine value is approximately 0.20113.

We can use the table feature of a calculator to find this angle, but it is faster to use the inverse sine key.

$$\sin^{-1} 0.20113 \approx \boxed{}°$$

Example 10   *Ladder Safety.* A paint crew has purchased new 30-ft extension ladders. The manufacturer states that the safest placement on a wall is to extend the ladder to 25 ft and to position the base 6.5 ft from the wall. What angle does the ladder make with the ground in this position?

We make a drawing to illustrate the situation.

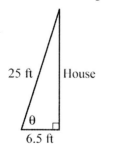

We know the length of the side adjacent to θ and the length of the hypotenuse, so we use the cosine function.

$$\cos\theta = \dfrac{6.5}{\boxed{}} = 0.26$$

We use the inverse cosine key on a calculator to find θ.
$$\theta \approx 75°$$

The angle that the ladder makes with the ground is approximately $\boxed{}°$.

Cofunction Identities

$\sin \theta = \cos(90° - \theta), \quad \cos \theta = \sin(90° - \theta),$

$\tan \theta = \cot(90° - \theta), \quad \cot \theta = \tan(90° - \theta),$

$\sec \theta = \csc(90° - \theta), \quad \csc \theta = \sec(90° - \theta)$

Example 11 Given that $\sin 18° \approx 0.3090,\ \cos 18° \approx 0.9511,$ and $\tan 18° \approx 0.3249,$ find the six trigonometric function values of $72°$.

Note that $18°$ and $\boxed{}°$ are complementary angles, so we can use the cofunction identities.

First find the three trigonometric function values of $18°$ that aren't given.

$$\csc 18° = \frac{1}{\sin 18°} \approx 3.2361, \quad \sec 18° = \frac{1}{\boxed{}18°} \approx 1.0515, \quad \cot 18° = \frac{1}{\tan\boxed{}°} \approx 3.0777$$

Now we find the six trigonometric function values of $72°$.

$$\sin 72° = \cos 18° \approx 0.9511$$

$$\cos 72° = \boxed{}18° \approx 0.3090$$

$$\tan 72° = \cot\boxed{}° \approx 3.0777$$

$$\cot 72° = \tan 18° \approx \boxed{}$$

$$\sec 72° = \csc 18° \approx 3.2361$$

$$\csc 72° = \sec\boxed{}° \approx 1.0515$$

Section 6.2   Applications of Right Triangles

To **solve** a triangle means to find the lengths of *all* the sides and the measures of *all* the angles.

Example 1   In $\triangle ABC$, find a, b, and B.

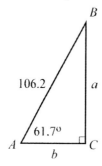

We know that $A = 61.7°$, $C = \boxed{}°$, and $c = 106.2$.

$$B = 90° - \boxed{}° = 28.3°$$

$$\sin 61.7° = \dfrac{a}{\boxed{}}$$

$$106.2 \sin 61.7° = \boxed{}$$

$$93.5 \approx a$$

$$\cos 61.7° = \dfrac{\boxed{}}{106.2}$$

$$106.2 \cos 61.7° = b$$

$$\boxed{} \approx b$$

Thus, we have the following

$$A = 61.7° \qquad a \approx \boxed{}$$

$$B = \boxed{}° \qquad b \approx 50.3$$

$$C = 90° \qquad c = \boxed{}$$

Example 2   In $\triangle DEF$, find D and F. Then find d.

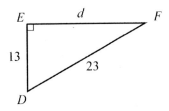

We know that $E = 90°$, $e = 23$, and $f = 13$.

$$\cos D = \frac{13}{\boxed{}}$$

$D \approx 55.58°$ Using the inverse cosine function on a calculator

$$F \approx 90° - 55.58° \approx \boxed{}°$$

$$\tan 55.58° = \frac{d}{\boxed{}}$$

$$13 \tan 55.58° = \boxed{}$$

$$19 \approx d$$

Thus, we have the following.

$$D \approx \boxed{}° \quad d = \boxed{}$$
$$E = 90° \qquad e = 23$$
$$F \approx 34.42° \quad f = \boxed{}$$

Example 3   *Walking at Niagara Falls.* While visiting Niagara Falls, a tourist walking toward Horseshoe Falls on a walkway next to Niagara Parkway notices the entrance to the Cave of the Winds attraction directly across the Niagara River. She continues walking for another 1000 ft and finds that the entrance is still visible but at approximately a $50°$ angle to the walkway.

a)   How many feet is she from the entrance to the Cave of the Winds?

b)   What is the approximate width of the Niagara River at that point?

We make a sketch.

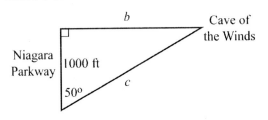

a) $\cos 50° = \dfrac{\boxed{}\ \text{ft}}{c}$

$\quad c = \dfrac{1000\ \text{ft}}{\cos 50°}$

$\quad c \approx \boxed{}\ \text{ft}$

After walking 1000 ft, she is approximately $\boxed{}$ ft from the entrance to the Cave of the Winds.

b) $\tan 50° = \dfrac{b}{\boxed{}\ \text{ft}}$

$\quad b = 1000\ \text{ft} \cdot \tan 50°$

$\quad b \approx \boxed{}\ \text{ft}$

The width of the Niagara River is approximately $\boxed{}$ ft at that point.

Example 4   *Rafters for a House.* House framers can use trigonometric functions to determine the lengths of rafters for a house. They first choose the pitch of the roof, or the ratio of the rise over the run. Then using a triangle with that ratio, they calculate the length of the rafter needed for the house. Jose is constructing rafters for a roof with a $10/12$ pitch on a house that is 42 ft wide. Find the length x of the rafter of the house to the nearest tenth of a foot.

We make a drawing.

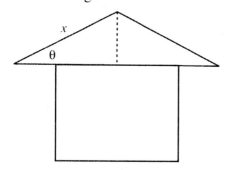

The pitch is the ratio $10/12$, so we draw a right triangle with sides in that ratio.

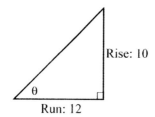

$$\tan \theta = \frac{10}{12} \approx 0.8333$$

$$\theta \approx \boxed{}{}^{\circ} \qquad\qquad \text{Using a calculator}$$

The width of the house (at the roof) is $\boxed{}$ ft, so in the right triangle with angle θ and hypotenuse x in the first drawing above, the side adjacent to θ is half of 42 ft or $\boxed{}$ ft.

$$\cos 39.8^{\circ} = \frac{21\ \text{ft}}{\boxed{}}$$

$$x \cos 39.8^{\circ} = 21\ \text{ft}$$

$$x = \frac{21\ \text{ft}}{\cos 39.8^{\circ}}$$

$$x \approx \boxed{}\ \text{ft}$$

The length of the rafter for this house is approximately 27.3 ft.

The angle between the horizontal and a line of sight above the horizontal is called an **angle of elevation**. The angle between the horizontal and a line of sight below the horizontal is called an **angle of depression**.

Example 5 *Gondola Aerial Lift.* In Telluride, Colorado, there is a free gondola ride that provides a spectacular view of the town and the surrounding mountains. The gondolas that begin in the town at an elevation of 8725 ft travel 5750 ft to Station St. Sophia, whose altitude is 10,550 ft. They then continue 3913 ft to Mountain Village, whose elevation is 9500 ft.

a)   What is the angle of elevation from the town to Station St. Sophia?

b)   What is the angle of depression from Station St. Sophia to Mountain Village?

a)   We first find the difference in the elevation of Station St. Sophia and the elevation of the town.

$$10{,}550\ \text{ft} - \boxed{}\ \text{ft} = 1825\ \text{ft}$$

Now we make a drawing, letting $\theta =$ the angle of elevation.

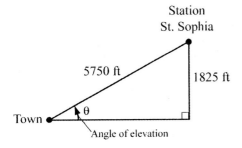

$$\sin \theta = \frac{1825 \text{ ft}}{\boxed{} \text{ ft}} \approx 0.3714$$

$$\theta \approx \boxed{}° \qquad\qquad \text{Using a calculator}$$

b)   The difference in the elevation of Station St. Sophia and the elevation of Mountain Village is

$$10,550 \text{ ft} - \boxed{} \text{ ft} = 1050 \text{ ft}.$$

We make a drawing, letting $\theta =$ the angle of depression.

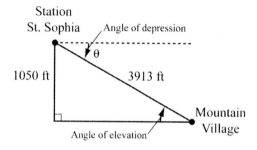

When parallel lines are cut by a transversal, alternate interior angles are equal. Thus the angle of depression θ, from Station St. Sophia to Mountain Village is equal to the angle of elevation from Mountain Village to Station St. Sophia.

$$\sin \theta = \frac{\boxed{} \text{ ft}}{3913 \text{ ft}} \approx 0.2683$$

$$\theta \approx \boxed{}° \qquad\qquad \text{Using a calculator}$$

Example 6   *Height of a Bamboo Plant.* Bamboo is the fastest growing land plant in the world and is becoming a popular wood for hardwood flooring. It can grow up to 46 in. per day and reaches its maximum height and girth in one season of growth. (*Source: Farm Show*, Vol. 34, Nov. 4, 2010, p. 7; *U-Cut Bamboo Business*; American Bamboo Society) To estimate the height of a bamboo shoot, a farmer walks off 27 ft from the base and estimates the angle of elevation to the top of the shoot to be 70°. What is the approximate height h of the bamboo shoot?

We make a drawing.

$$\tan 70° = \frac{\boxed{}}{27 \text{ ft}}$$

$$h = 27 \text{ ft} \cdot \tan 70° \approx \boxed{} \text{ ft}$$

The height of the bamboo shoot is approximately 74 ft.

One method of giving direction, or **bearing**, involves reference to a north-south line using an acute angle. For example, N55°W means 55° west of north and S67°E means 67° east of south.

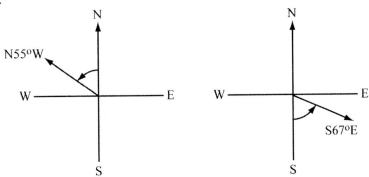

Example 7   A forest ranger at point A sights a fire directly south. A second ranger at point B, 7.5 mi east of the first ranger, sights the same fire at a bearing of S27°23'W. How far from A is the fire?

We make a drawing. Note that $B = 90° - 27°23' = 62°37' = 62\dfrac{37}{60}^{\circ} \approx 62.62°$.

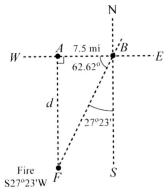

$$\tan 62°37' \approx \tan 62.62° = \frac{\square}{7.5 \text{ mi}}$$

$$7.5 \text{ mi} \cdot \tan 62.62° \approx d$$

$$\boxed{} \text{ mi} \approx d$$

The forest ranger at A is about 14.5 mi from the fire.

Example 8   *U.S. Cellular Field.* In U.S. Cellular Field, the home of the Chicago White Sox baseball team, the first row of seats in the upper deck is farther away from home plate than the last row of seats in the original Comiskey Park. Although there is no obstructed view in U.S. Cellular Field, some of the fans still complain about the distance from home plate to the upper deck of seats. From a seat in the last row of the upper deck directly behind the batter, the angle of depression to home plate is $29.9°$, and the angle of depression to the pitcher's mound is $24.2°$. Find (a) the viewing distance to home plate, and (b) the viewing distance to the pitcher's mound.

We make a drawing. Note that the standard distance from the pitcher's mound to home plate is 60.5 ft.

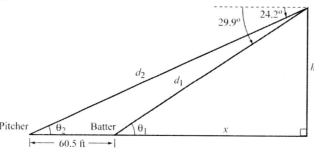

From geometry we know that $\theta_1 = 29.9°$ and $\theta_2 = 24.2°$.

$$\tan 29.9° = \frac{\square}{x}, \quad \tan 24.2° = \frac{\square}{x+60.5}$$

$$h = x \tan 29.9°, \quad h = (x+60.5)\tan\boxed{}°$$

Substitute $x \tan 29.9°$ for h in the second equation.

$$x \tan 29.9° = \left(x + \boxed{}\right)\tan 24.2°$$

$$x \tan 29.9° = x \tan 24.2° + 60.5 \tan 24.2°$$

$$x \tan 29.9° - x \tan 24.2° = 60.5 \tan 24.2°$$

$$x\left(\tan 29.9° - \tan 24.2°\right) = \boxed{}\tan 24.2°$$

$$x = \frac{60.5 \tan 24.2°}{\tan 29.9° - \tan\boxed{}°}$$

$$x \approx 216.5$$

We use the value of x to find d_1 and d_2.

$$\cos 29.9° = \frac{216.5}{\boxed{}} \quad \text{and} \quad \cos 24.2° = \frac{60.5 + 216.5}{\boxed{}}$$

$$d_1 = \frac{216.5}{\cos 29.9°} \quad \text{and} \quad d_2 = \frac{277}{\cos 24.2°}$$

$$d_1 \approx 249.7 \text{ ft} \quad \text{and} \quad d_2 \approx \boxed{} \text{ ft}$$

Section 6.3   Trigonometric Functions of Any Angle

If two or more angles have the same terminal side, the angles are said to be **coterminal**.

Example 1   Find two positive angles and two negative angles that are coterminal with (a) $51°$ and (b) $-7°$.

a) To find positive angles that are coterminal with $51°$, we add multiples of $360°$.

$$51° + 360° = \boxed{}°$$
$$51° + 3(360°) = \boxed{}°$$

To find negative angles that are coterminal with $51°$, we subtract multiples of $360°$.

$$51° - 360° = \boxed{}°$$
$$51° - 2(360°) = \boxed{}°$$

b) $-7° + 360° = 353°$

$$-7° + 2(360°) = -7° + \boxed{}° = 713°$$
$$-7° - 360° = \boxed{}°$$
$$-7° - 10(360°) = -7° - \boxed{}° = -3607°$$

Example 2   Find the complement and the supplement of $71.46°$.

Two acute angles are complementary if the sum of their measures is $90°$. The complement is

$$\boxed{}° - 71.46° = 18.54°.$$

Two positive angles are supplementary if the sum of their measures is $180°$. The supplement is

$$180° - \boxed{}° = 108.54°.$$

> ### Trigonometric Functions of Any Angle θ
>
> Suppose that $P(x, y)$ is any point other than the vertex on the terminal side of any angle θ in standard position, and r is the radius, or distance from the origin to $P(x, y)$. Then the trigonometric functions are defined as follows:
>
> $$\sin\theta = \frac{y\text{-coordinate}}{\text{radius}} = \frac{y}{r}, \qquad\qquad \csc\theta = \frac{\text{radius}}{y\text{-coordinate}} = \frac{r}{y} \ (y \neq 0),$$
>
> $$\cos\theta = \frac{x\text{-coordinate}}{\text{radius}} = \frac{x}{r}, \qquad\qquad \sec\theta = \frac{\text{radius}}{x\text{-coordinate}} = \frac{r}{x} \ (x \neq 0),$$
>
> $$\tan\theta = \frac{y\text{-coordinate}}{x\text{-coordinate}} = \frac{y}{x} \ (x \neq 0), \qquad \cot\theta = \frac{x\text{-coordinate}}{y\text{-coordinate}} = \frac{x}{y} \ (y \neq 0).$$

Example 3   Find the six trigonometric function values for each angle shown.

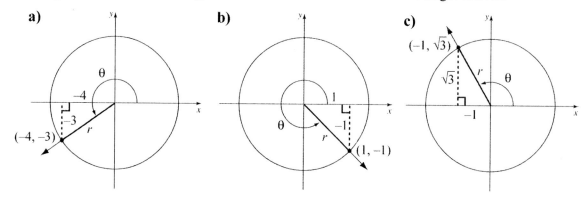

a)   Distance from $(0,0)$ to $(-4, -3)$:

$$r = \sqrt{(-4-0)^2 + (-3-0)^2}$$

$$= \sqrt{(-4)^2 + (-3)^2}$$

$$= \sqrt{\boxed{}+9} = \sqrt{\boxed{}} = 5.$$

We have $x = -4$, $y = -3$, $r = 5$.

$$\sin\theta = \frac{y}{r} = \frac{-3}{\boxed{}} = -\frac{3}{5}, \quad \cos\theta = \frac{x}{r} = \frac{\boxed{}}{5} = -\frac{4}{5}, \quad \tan\theta = \frac{y}{x} = \frac{-3}{\boxed{}} = \frac{3}{4},$$

$$\csc\theta = \frac{r}{y} = \frac{\boxed{}}{-3} = -\frac{5}{3}, \quad \sec\theta = \frac{r}{x} = \frac{5}{\boxed{}} = -\frac{5}{4}, \quad \cot\theta = \frac{x}{y} = \frac{\boxed{}}{-3} = \frac{4}{3}$$

b) Distance from $(0, 0)$ to $(1, -1)$:

$$r = \sqrt{(1-0)^2 + (-1-0)^2}$$
$$= \sqrt{1^2 + (-1)^2}$$
$$= \sqrt{1 + \square} = \sqrt{2}.$$

We have $x = 1$, $y = -1$, $r = \sqrt{2}$.

$$\sin\theta = \frac{y}{r} = \frac{\square}{\sqrt{2}} = -\frac{\sqrt{2}}{2}, \quad \cos\theta = \frac{x}{r} = \frac{\square}{\sqrt{2}} = \frac{\sqrt{2}}{2}, \quad \tan\theta = \frac{y}{x} = \frac{\square}{1} = -1,$$

$$\csc\theta = \frac{r}{y} = \frac{\sqrt{2}}{\square} = \square, \quad \sec\theta = \frac{r}{x} = \frac{\sqrt{2}}{\square} = \square, \quad \cot\theta = \frac{x}{y} = \frac{\square}{-1} = -1$$

c) Distance from $(0, 0)$ to $\left(-1, \sqrt{3}\right)$:

$$r = \sqrt{(-1-0)^2 + \left(\sqrt{3}-0\right)^2}$$
$$= \sqrt{(-1)^2 + \left(\sqrt{3}\right)^2}$$
$$= \sqrt{1+3} = \sqrt{\square} = \square.$$

We have $x = -1$, $y = \sqrt{3}$, $r = 2$.

$$\sin\theta = \frac{y}{r} = \frac{\sqrt{3}}{\square}, \quad \cos\theta = \frac{x}{r} = \frac{-1}{\square} = -\frac{1}{2}, \quad \tan\theta = \frac{y}{x} = \frac{\sqrt{3}}{\square} = \square,$$

$$\csc\theta = \frac{r}{y} = \frac{\square}{\sqrt{3}} = \frac{2\sqrt{3}}{3}, \quad \sec\theta = \frac{r}{x} = \frac{2}{\square} = \square, \quad \cot\theta = \frac{x}{y} = \frac{\square}{\sqrt{3}} = -\frac{\sqrt{3}}{3}$$

Example 4 Given that $\tan\theta = -\dfrac{2}{3}$ and θ is in the second quadrant, find the other function values.

$\tan\theta = \dfrac{y}{x}$; θ is in quadrant II, so x is negative and y is positive. We have

$$\tan\theta = -\frac{2}{3} = \frac{2}{-3}.$$

Distance from $(0, 0)$ to $(-3, 2)$:

$$r = \sqrt{(-3-0)^2 + \left(\square - 0\right)^2}$$
$$= \sqrt{(-3)^2 + 2^2} = \sqrt{9+4} = \sqrt{\square}.$$

We have $x = -3$, $y = 2$, $r = \sqrt{13}$.

$$\sin\theta = \frac{y}{r} = \frac{2}{\sqrt{13}} = \frac{2\sqrt{13}}{\boxed{}}, \quad \cos\theta = \frac{x}{r} = \frac{\boxed{}}{\sqrt{13}} = -\frac{3\sqrt{13}}{13},$$

$$\tan\theta = -\frac{2}{3} \quad \text{(We were given this.)}$$

$$\csc\theta = \frac{r}{y} = \frac{\sqrt{13}}{\boxed{}}, \quad \sec\theta = \frac{r}{x} = \frac{\sqrt{13}}{\boxed{}} = -\frac{\sqrt{13}}{3}, \quad \cot\theta = \frac{x}{y} = \frac{-3}{\boxed{}} = -\frac{3}{2}$$

Example 5   Find the sine, cosine, and tangent values for 90°, 180°, 270°, and 360°.

For 90°:

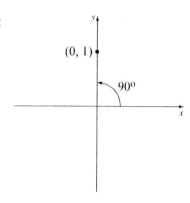

$x = 0$, $y = 1$, $r = 1$

$$\sin 90° = \frac{1}{1} = \boxed{}$$

$$\cos 90° = \frac{0}{1} = 0$$

$$\tan 90° = \frac{1}{\boxed{}} \quad \text{This is not defined.}$$

For 180°:

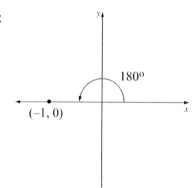

$x = -1$, $y = 0$, $r = 1$

$$\sin 180° = \frac{\boxed{}}{1} = 0$$

$$\cos 180° = \frac{-1}{1} = \boxed{}$$

$$\tan 180° = \frac{0}{\boxed{}} = 0$$

For 270°:

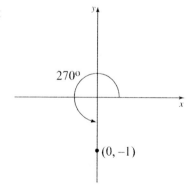

$x = 0$, $y = -1$, $r = 1$

$$\sin 270° = \frac{-1}{1} = -1$$

$$\cos 270° = \frac{0}{1} = \boxed{}$$

$$\tan 270° = \frac{-1}{\boxed{}} \quad \text{This is not defined.}$$

For $360°$:

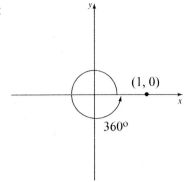

(1, 0)

$360°$

$x = 1,\ y = 0,\ r = 1$

$$\sin 360° = \frac{\boxed{}}{1} = 0$$

$$\cos 360° = \frac{1}{\boxed{}} = 1$$

$$\tan 360° = \frac{\boxed{}}{1} = 0$$

Example 6 Find each of the following:

a) $\sin(-90°)$ b) $\csc 540°$

a) $-90°$ is coterminal with $270°$.

$$\sin(-90°) = \sin 270° = \boxed{}$$

b) Since $540° = 180° + 360°$, $540°$ and $180°$ are coterminal.

$$\csc 540° = \csc 180° = \frac{1}{\sin 180°} = \frac{1}{\boxed{}}, \text{ which is not defined.}$$

Reference Angle

The **reference angle** for an angle is the acute angle formed by the terminal side of the angle and the x-axis.

Example 7 Find the sine, cosine, and tangent function values for each of the following:

a) $225°$ b) $-780°$

a) $225°$ is a third quadrant angle. The reference angle is $225° - 180° = \boxed{}°$.

Recall that $\sin 45° = \sqrt{2}/2$, $\cos 45° = \sqrt{2}/2$, and $\tan 45° = 1$. Also note that in the third quadrant, the sine and the cosine are negative and the tangent is positive.

$$\sin 225° = -\frac{\sqrt{2}}{\boxed{}}, \quad \cos 225° = -\frac{\sqrt{2}}{2}, \quad \tan 225° = \boxed{}$$

b) $-780° = 2(-360°) + (-60°)$, so $-780°$ and $-60°$ are coterminal, and they are fourth quadrant angles. The reference angle for $-60°$ is $\boxed{}°$.

Recall that $\sin 60° = \sqrt{3}/2$, $\cos 60° = 1/2$, and $\tan 60° = \sqrt{3}$. In quadrant IV, the

cosine is $\underline{\hspace{3cm}}$ and the sine and tangent are $\underline{\hspace{3cm}}$.
$\quad\quad\quad\quad$ positive / negative $\quad\quad\quad\quad\quad\quad\quad\quad\quad$ positive / negative

$$\sin(-780°) = -\frac{\sqrt{3}}{\boxed{}}, \quad \cos(-780°) = \frac{\boxed{}}{2}, \quad \tan(-780°) = \boxed{}$$

Example 8   Find each of the following function values using a calculator and round the answer to four decimal places, where appropriate.

a) $\cos 112°$ b) $\sec 500°$

c) $\tan(-83.4°)$ d) $\csc 351.75°$

e) $\cos 2400°$ f) $\sin 175°40'9''$

g) $\cot(-135°)$

Using a calculator set in DEGREE mode, we find the values.

a) $\cos 112° \approx -0.3746$

b) $\sec 500° = \dfrac{1}{\cos 500°} \approx \boxed{}$

c) $\tan(-83.4°) \approx -8.6427$

d) $\csc 351.75° = \dfrac{1}{\boxed{}\,351.75°} \approx -6.9690$

e) $\cos 2400° = \boxed{}$

f) $\sin 175°40'9'' \approx 0.0755$

g) $\cot(-135°) = \dfrac{1}{\tan(-135°)} = \boxed{}$

Example 9   Given the function value and the quadrant restrictions, find θ.

a) $\sin\theta = 0.2812,\ 90° < \theta < 180°$

b) $\cot\theta = -0.1611,\ 270° < \theta < 360°$

a)   We make a drawing. θ is in quadrant II.

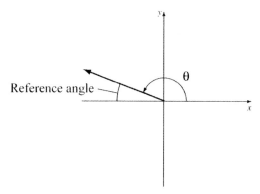

Using a calculator, we find that the reference angle is approximately $16.33°$.

$$\theta \approx \boxed{}° - 16.33° \approx 163.67°$$

b) We make a drawing. θ is in quadrant IV.

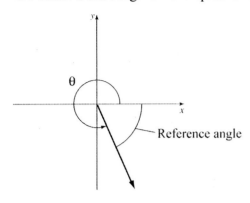

Reference angle

$$\tan\theta = \frac{1}{\cot\theta} = \frac{1}{\boxed{}} \approx -6.2073$$

Using a calculator, we find that the reference angle is approximately $80.85°$.

$$\theta \approx \boxed{}° - 80.85° \approx 279.15°$$

In aerial navigation, directions are given in degrees clockwise from north. Thus east is $90°$, south is $180°$, and west is $270°$. Two examples of aerial directions, or **bearings**, are shown below.

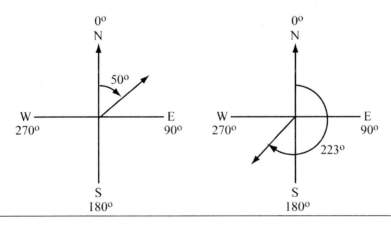

Example 10   *Aerial Navigation.*   An airplane flies 218 mi from an airport in a direction of 245°. How far south of the airport is the plane then? How far west?

We make a drawing.

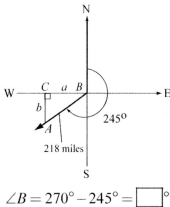

$$\angle B = 270° - 245° = \boxed{}°$$

$$\sin 25° = \frac{b}{218}$$

$$218 \sin 25° = \boxed{}$$

$$92 \approx b$$

The plane is approximately $\boxed{}$ mi south of the airport.

$$\cos 25° = \frac{a}{\boxed{}}$$

$$218 \cos 25° = a$$

$$\boxed{} \approx a$$

The plane is approximately 198 mi west of the airport.

Section 6.4   Radians, Arc Length, and Angular Speed

Example 1   How far will a point travel if it goes (a) $\dfrac{1}{4}$, (b) $\dfrac{1}{12}$, (c) $\dfrac{3}{8}$, and (d) $\dfrac{5}{6}$ of the way around the unit circle?

The circumference C of a circle with radius r is given by $C = 2\pi r$. In the unit circle $r = 1$, so the circumference is 2π.

a)   The distance is $\dfrac{1}{4} \cdot 2\square = \dfrac{1}{2}\pi$, or $\dfrac{\pi}{2}$.

b)   The distance is $\dfrac{1}{\square} \cdot 2\pi = \dfrac{\pi}{6}$.

c)   The distance is $\dfrac{\square}{8} \cdot 2\pi = \dfrac{6\pi}{8}$, or $\dfrac{3\pi}{\square}$.

d)   The distance is $\dfrac{5}{6} \cdot \square = \dfrac{10\pi}{6} = \dfrac{\square \pi}{3}$.

Example 2   On the unit circle, mark the point determined by each of the following real numbers.

a) $\dfrac{9\pi}{4}$

b) $-\dfrac{7\pi}{6}$

a) $\dfrac{9\pi}{4}$ is positive, so we will move _____ around the unit circle.
clockwise / counterclockwise

$$\dfrac{9\pi}{4} = \dfrac{8\pi}{4} + \dfrac{\square}{4} = 2\pi + \dfrac{1}{4}\pi$$

From the point A in the figure below, move around the circle once and then continue $\dfrac{1}{4}$ of the way from A to B. Mark this point.

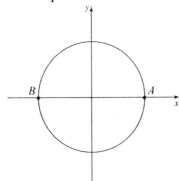

b) $-\dfrac{7\pi}{6}$ is negative, so we will move _____ around the unit

                                                 clockwise / counterclockwise

circle.

$$-\frac{7\pi}{6} = -\frac{6}{6}\pi + \left(-\frac{1}{6}\pi\right) = -\pi + \left(-\frac{1}{6}\pi\right)$$

We move from A to B and then beyond B another distance of $\dfrac{\pi}{6}$. Mark this point.

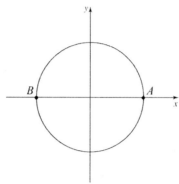

Converting Between Degree Measure and Radian Measure

$$\frac{\pi \text{ radians}}{180°} = \frac{180°}{\pi \text{ radians}} = 1.$$

To convert from degree to radian measure, multiply by $\dfrac{\pi \text{ radians}}{180°}$.

To convert from radian to degree measure, multiply by $\dfrac{180°}{\pi \text{ radians}}$.

Example 3   Convert each of the following to radians.

a) $120°$ b) $-297.25°$

a) $120° = 120° \cdot 1 = 120° \cdot \dfrac{\pi \text{ radians}}{\boxed{}°}$

                  $= \dfrac{120°}{180°} \cdot \pi \text{ radians}$

                  $= \dfrac{2\pi}{\boxed{}} \text{ radians} \approx 2.09 \text{ radians}$

b) $-297.25° = -297.25° \cdot 1 = -297.25° \cdot \dfrac{\boxed{} \text{ radians}}{180°}$

                        $= -\dfrac{297.25° \cdot \pi}{\boxed{}°} \text{ radians}$

                        $\approx -5.19 \text{ radians}$

Example 4   Convert each of the following to degrees.

a) $\dfrac{3\pi}{4}$ radians

b)   8.5 radians

a) $\dfrac{3\pi}{4}$ radians $= \dfrac{3\pi}{4}$ radians $\cdot \dfrac{\boxed{}^{\circ}}{\pi \text{ radians}}$

$ = \dfrac{3\pi \cdot 180^{\circ}}{4\pi}$

$ = \dfrac{3 \cdot 180^{\circ}}{\boxed{}} = 135^{\circ}$

b)   8.5 radians $= 8.5$ radians $\cdot \dfrac{180^{\circ}}{\boxed{} \text{ radians}}$

$ = \dfrac{8.5\left(180^{\circ}\right)}{\boxed{}} \approx 487.01^{\circ}$

Example 5   Find a positive angle and a negative angle that are coterminal with $2\pi/3$. Many answers are possible.

We add or subtract multiples of 2π.

$$\dfrac{2\pi}{3} + 2\pi = \dfrac{2\pi}{3} + \dfrac{6\pi}{\boxed{}} = \dfrac{\boxed{}\pi}{3}$$

$$\dfrac{2\pi}{3} - 3\left(2\pi\right) = \dfrac{2\pi}{3} - 6\pi = \dfrac{2\pi}{3} - \dfrac{18\pi}{\boxed{}} = -\dfrac{16\boxed{}}{3}$$

Example 6   Find the complement and the supplement of $\pi/6$.

$90^{\circ} = \dfrac{\pi}{2}$, so the complement of $\dfrac{\pi}{6}$ is

$$\dfrac{\pi}{2} - \dfrac{\pi}{6} = \dfrac{3\pi}{\boxed{}} - \dfrac{\pi}{6} = \dfrac{2\pi}{6} = \dfrac{\boxed{}}{3}.$$

$180^{\circ} = \pi$, so the supplement of $\dfrac{\pi}{6}$ is

$$\pi - \dfrac{\pi}{6} = \dfrac{6\pi}{\boxed{}} - \dfrac{\pi}{6} = \dfrac{\boxed{}\pi}{6}.$$

Radian Measure

The **radian measure** θ of a rotation is the **ratio** of the distance s, traveled by a point at a radius r from the center of rotation, to the length of the radius r.

$\theta = \dfrac{s}{r}$, where θ is in radians and s and r must be expressed in the same unit.

Example 7   Find the measure of a rotation in radians when a point 2 m from the center of rotation travels 4 m.

$$\theta = \frac{s}{r} = \frac{4 \text{ m}}{\boxed{} \text{ m}} = 2$$

Example 8   Find the length of an arc of a circle of radius 5 cm associated with an angle of $\pi / 3$ radians.

$$\theta = \frac{s}{r}, \text{ or } s = r\theta.$$

$$s = \boxed{} \text{ cm} \cdot \frac{\pi}{3} \approx 5.24 \boxed{}$$

Linear Speed in Terms of Angular Speed

The **linear speed** v of a point a distance r from the center of rotation is given by

$$v = r\omega,$$

where ω is the **angular speed**, in radians per unit of time.

The units of distance for v and r must be the same, ω must be in radians per unit of time, and the units of time for v and ω must be the same.

Example 9   *Linear Speed of an Earth Satellite.* An earth satellite in circular orbit 1200 km high makes one complete revolution every 90 min. What is its linear speed? Use 6400 km for the length of a radius of the earth.

We will use the formula $v = r\omega$. We first find r and ω.

$$\omega = \frac{\theta}{t} = \frac{2\pi}{90 \text{ min}} = \frac{\pi}{\boxed{} \text{ min}}$$

Now substitute in the formula for linear speed.

$$v = r\omega = 7600 \text{ km} \cdot \frac{\pi}{45 \text{ min}} \qquad \left(1200 \text{ km} + 6400 \text{ km} = 7600 \text{ km}\right)$$

$$= \frac{7600\pi}{\boxed{}} \cdot \frac{\text{km}}{\text{min}} \approx 531 \frac{\text{km}}{\text{min}}$$

Example 10 *Angular Speed of a Capstan.* An anchor on a Navy vessel is hoisted at a rate of 2 ft/sec as the line is wound around a capstan with 1.4-yd diameter. What is the angular speed of the capstan?

We will use the formula $v = r\omega$. First we find r and v. The units of distance must be the same for v and r. We will use feet for both and also use the diameter given to find the radius of the capstan.

$$r = \frac{d}{2} = \frac{1.4 \text{ yd}}{2} = 0.7 \text{ yd} \times \frac{\boxed{} \text{ ft}}{1 \text{ yd}} = 2.1 \,\boxed{}$$

$v = r\omega$, or $\omega = \dfrac{v}{r}$.

$$\omega = \frac{2 \text{ ft/sec}}{\boxed{} \text{ ft}} = \frac{2 \text{ ft}}{1 \text{ sec}} \cdot \frac{1}{2.1 \text{ ft}} \approx \boxed{} / \text{sec}$$

The angular speed is approximately 0.952 radians per second.

Example 11 *Angle of Revolution.* A 2014 Toyota FJ Cruiser is traveling at a speed of 70 mph. Its tires have an outside diameter of 30.875 in. Find the angle through which a tire turns in 10 sec.

We will use the formula $\omega = \dfrac{\theta}{t}$ in the form $\theta = \omega t$. The distance and time units must be the same. We will convert 70 mph to feet per second.

$$v = \frac{70 \text{ mi}}{1 \text{ hr}} \cdot \frac{1 \text{ hr}}{\boxed{} \text{ min}} \cdot \frac{1 \text{ min}}{60 \text{ sec}} \cdot \frac{5280 \text{ ft}}{\boxed{} \text{ mi}} \approx 102.667 \,\frac{\text{ft}}{\text{sec}}$$

Now we use $v = r\omega$ to find ω. First we will find the radius of the tire and then convert it to ft.

$$r = \frac{d}{2} = \frac{30.875 \text{ in.}}{2} \approx 15.4375 \text{ in.} \cdot \frac{1 \text{ ft}}{\boxed{} \text{ in.}} \approx 1.29 \,\boxed{}$$

Now we find ω.

$$v = r\omega$$

$$102.667 \,\frac{\text{ft}}{\text{sec}} = 1.29 \text{ ft} \cdot \omega$$

$$\frac{102.667 \text{ ft/sec}}{1.29 \text{ ft}} \approx \omega$$

$$79.59 / \text{sec} \approx \omega$$

Finally, we find θ. We have $\omega = \dfrac{\theta}{t}$, so $\omega \cdot t = \theta$.

$$\theta = \omega t \approx \frac{79.59}{\text{sec}} \cdot \boxed{} \text{ sec} \approx 796$$

The angle in radians, through which a tire turns in 10 sec is 796.

Example 12   One gear wheel turns another, the teeth being on the rims. The wheels have 9-in. and 5-in. radii, and the smaller wheel rotates at 48 rpm. Find the angular speed of the larger wheel, in radians per second.

Let $\omega_1 =$ the angular speed of the 5-in. wheel, and let $\omega_2 =$ the angular speed of the 9-in. wheel. The wheels have the same linear speed so we have

$$v = 5\omega_1 \text{ and } v = 9\omega_2, \text{ or } 5\omega_1 = 9\omega_2.$$

Now we convert 48 rpm to radians per second

$$\omega_1 = 48 \text{ rpm} = 48 \cdot \frac{2\pi}{1 \text{ min}} = \frac{96\pi}{1 \text{ min}} \cdot \frac{1 \text{ min}}{\boxed{} \text{ sec}} = 1.6\,\pi\,/\,\text{sec}$$

Now we find ω_2.

$$5\omega_1 = 9\omega_2$$
$$5(1.6\,\pi\,/\,\text{sec}) = 9\omega_2$$
$$\frac{5(1.6\,\pi\,/\,\text{sec})}{9} = \omega_2$$
$$2.793\,/\,\text{sec} \approx \omega_2$$

Section 6.5   Circular Functions: Graphs and Properties

Basic Circular Functions

For a real number s that determines a point (x, y) on the unit circle:

$$\sin s = \text{second coordinate} = \frac{y}{1} = y,$$

$$\cos s = \text{first coordinate} = \frac{x}{1} = x,$$

$$\tan s = \frac{\text{second coordinate}}{\text{first coordinate}} = \frac{y}{x} \quad (x \neq 0),$$

$$\csc s = \frac{1}{\text{second coordinate}} = \frac{1}{y} \quad (y \neq 0),$$

$$\sec s = \frac{1}{\text{first coordinate}} = \frac{1}{x} \quad (x \neq 0),$$

$$\cot s = \frac{\text{first coordinate}}{\text{second coordinate}} = \frac{x}{y} \quad (y \neq 0).$$

Example 1   Each of the following points lies on the unit circle. Find their reflections across the x-axis, the y-axis, and the origin.

a) $\left(\dfrac{3}{5}, \dfrac{4}{5}\right)$
b) $\left(\dfrac{\sqrt{2}}{2}, \dfrac{\sqrt{2}}{2}\right)$
c) $\left(\dfrac{1}{2}, \dfrac{\sqrt{3}}{2}\right)$

a)

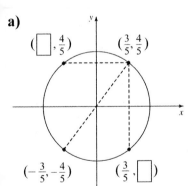

b)

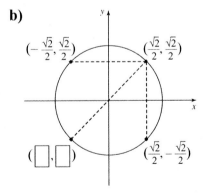

c)

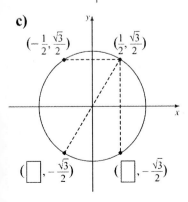

Example 2 Find each of the following function values.

a) $\tan\dfrac{\pi}{3}$

b) $\cos\dfrac{3\pi}{4}$

c) $\sin\left(-\dfrac{\pi}{6}\right)$

d) $\cos\dfrac{4\pi}{3}$

e) $\cot\pi$

f) $\csc\left(-\dfrac{7\pi}{2}\right)$

We locate the point on the unit circle determined by the rotation and then find its coordinates using reflection, if necessary.

a)

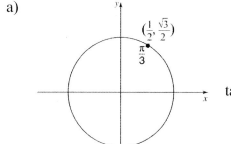

$$\tan\frac{\pi}{3} = \frac{y}{\square} = \frac{\sqrt{3}/2}{1/2}$$

$$= \frac{\sqrt{3}}{2}\cdot\frac{2}{1} = \square$$

b) The coordinates of the point determined by $\dfrac{\pi}{4}$ are $\left(\dfrac{\sqrt{2}}{2},\dfrac{\sqrt{2}}{2}\right)$. The rotation $\dfrac{3\pi}{4}$ is the

reflection of $\dfrac{\pi}{4}$ across the y-axis. Thus the coordinates of the point determined by $\dfrac{3\pi}{4}$

are $\left(\square,\dfrac{\sqrt{2}}{2}\right)$.

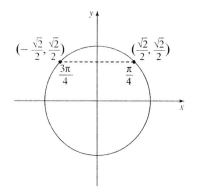

$$\cos\frac{3\pi}{4} = x = \square$$

c) The coordinates of the point determined by $\dfrac{\pi}{6}$ are $\left(\dfrac{\sqrt{3}}{2},\dfrac{1}{2}\right)$. The rotation $-\dfrac{\pi}{6}$ is a

reflection of $\dfrac{\pi}{6}$ across the _____-axis. Thus, the coordinates of the point determined by
$$\underset{x/y}{}$$

$-\dfrac{\pi}{6}$ are $\left(\dfrac{\sqrt{3}}{2},\square\right)$.

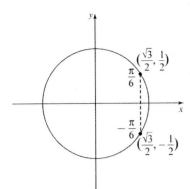

$$\sin\left(-\frac{\pi}{6}\right) = y = \boxed{}$$

d)   The coordinates of the point determined by $\frac{\pi}{3}$ are $\left(\frac{1}{2}, \frac{\sqrt{3}}{2}\right)$. The rotation $\frac{4\pi}{3}$ is a

reflection of $\frac{\pi}{3}$ across the origin. Thus, the coordinates of the point determined by $\frac{4\pi}{3}$

are $\left(\boxed{}, -\frac{\sqrt{3}}{2}\right)$.

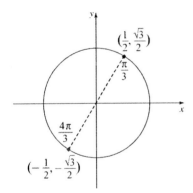

$$\cos\frac{4\pi}{3} = x = \boxed{}$$

e)

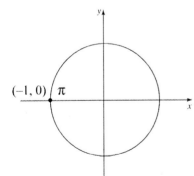

$$\cot \pi = \frac{x}{y} = \frac{-1}{\boxed{}}, \text{ which is not defined}.$$

f)

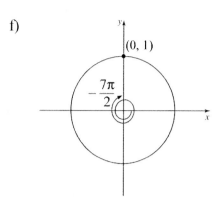

$$\csc\left(-\frac{7\pi}{2}\right) = \frac{1}{y} = \frac{1}{\boxed{}} = 1$$

Example 3   Find each of the following function values of radian measures using a calculator. Round the answer to four decimal places.

a) $\cos\dfrac{2\pi}{5}$

b) $\tan(-3)$

c) $\sin 24.9$

d) $\sec\dfrac{\pi}{7}$

Using a calculator set in RADIAN mode, we find the values.

a) $\cos\dfrac{2\pi}{5} \approx 0.3090$

b) $\tan(-3) \approx \boxed{}$

c) $\sin 24.9 \approx -0.2306$

d) $\sec\dfrac{\pi}{7} = \dfrac{1}{\boxed{}\dfrac{\pi}{7}} \approx 1.1099$

Section 6.6   Graphs of Transformed Sine Functions and Cosine Functions

---

The Constant _D_: Translating Vertically

The constant D in $y = A\sin(Bx - C) + D$ and $y = A\cos(Bx - C) + D$ *translates* the graphs *vertically* up D units if $D > 0$ or down $|D|$ units if $D < 0$.

Example 1   Sketch a graph of $y = \sin x + 3$.

This is a vertical translation of the graph of $y = \sin x$ _____ 3 units.

up / down

Consider the following key points on the graph of $y = \sin x$:

$$(0,0),\ \left(\frac{\pi}{2}, 1\right),\ (\pi, 0),\ \left(\frac{3\pi}{2}, -1\right),\ (2\pi, 0).$$

We increase each y-coordinate by $\boxed{}$ to obtain key points on the graph of $y = \sin x + 3$. We have

$$\left(0, \boxed{}\right),\ \left(\frac{\pi}{2}, 4\right),\ (\pi, 3),\ \left(\frac{3\pi}{2}, \boxed{}\right),\ \left(2\pi, \boxed{}\right).$$

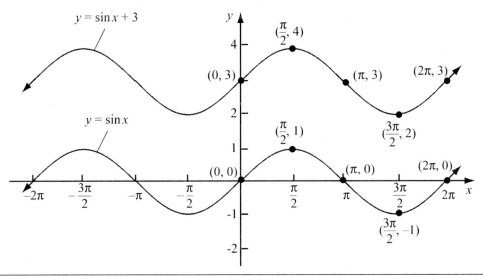

The Constant _A_: Vertical Stretching and Shrinking

The constant A in $y = A\sin(Bx - C) + D$ and $y = A\cos(Bx - C) + D$ *stretches* or *shrinks* the graph *vertically*. If $|A| > 1$, then there will be a vertical stretching. If $|A| < 1$, then there will be a vertical shrinking. If $A < 0$, the graph is also *reflected* across the x-axis.

The **amplitude** of the graphs of $y = A\sin(Bx - C) + D$ and $y = A\cos(Bx - C) + D$ is $|A|$.

Example 2   Sketch a graph of $y = 2\cos x$. What is the amplitude?

The amplitude is $|2| = 2$.

This is a vertical stretching of $y = \cos x$ by a factor of $\boxed{}$.

Consider the following key points on the graph of $y = \cos x$:

$$(0, 1),\ \left(\frac{\pi}{2}, 0\right),\ (\pi, -1),\ \left(\frac{3\pi}{2}, 0\right),\ (2\pi, 1).$$

We multiply each second coordinate by $\boxed{}$ to obtain key points on the graph of $y = 2\cos x$. We have

$$\left(0, \boxed{}\right),\ \left(\frac{\pi}{2}, 0\right),\ \left(\pi, \boxed{}\right),\ \left(\frac{3\pi}{2}, \boxed{}\right),\ (2\pi, 2).$$

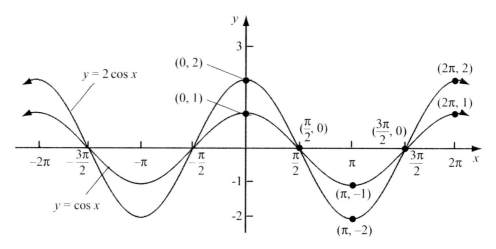

Example 3   Sketch a graph of $y = -\frac{1}{2}\sin x$. What is the amplitude?

The amplitude is $\left|-\frac{1}{2}\right| = \frac{1}{2}$.

Each y-coordinate on the graph of $y = \sin x$ will be multiplied by $-\frac{1}{2}$ to obtain the

corresponding point on the graph of $y = -\frac{1}{2}\sin x$. This will produce a vertical shrinking of

$y = \sin x$ as well as the reflection across the x-axis. The key points of $y = \sin x$,

$$(0, 0),\ \left(\frac{\pi}{2}, 1\right),\ (\pi, 0),\ \left(\frac{3\pi}{2}, -1\right),\ (2\pi, 0),$$

are transformed to

$$(0, 0),\ \left(\frac{\pi}{2}, \boxed{}\right),\ \left(\pi, \boxed{}\right),\ \left(\frac{3\pi}{2}, \boxed{}\right),\ \text{and } (2\pi, 0).$$

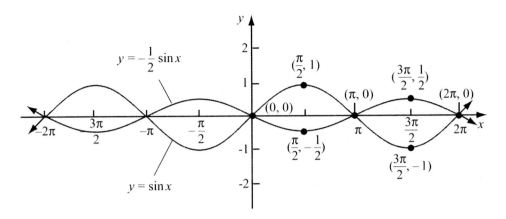

The Constant B: Stretching or Shrinking Horizontally

The constant B in $y = A \sin(Bx - C) + D$ and $y = A \cos(Bx - C) + D$ *stretches* or *shrinks* the graph *horizontally*. If $|B| < 1$, then there will be a horizontal stretching. If $|B| > 1$, then there will be a horizontal shrinking. If $B < 0$, the graph is also *reflected* across the y-axis.

The **period** of the graphs of $y = A \sin(Bx - C) + D$ and $y = A \cos(Bx - C) + D$ is $\left| \dfrac{2\pi}{B} \right|$.

Example 4   Sketch a graph of $y = \sin 4x$. What is the period?

The period is $\left| \dfrac{2\pi}{4} \right| = \left| \dfrac{\pi}{2} \right| = \dfrac{\pi}{2}$.

This is a horizontal _____ of the graph of $y = \sin x$. Each first coordinate
 stretching / shrinking

on the graph of $y = \sin x$ will be divided by $\boxed{}$ to obtain the corresponding point on the graph of $y = \sin 4x$. Consider the key points of $y = \sin x$:

$$(0,0), \ \left(\frac{\pi}{2}, 1\right), \ (\pi, 0), \ \left(\frac{3\pi}{2}, -1\right), \ (2\pi, 0).$$

They are transformed to

$$\left(\boxed{}, 0\right), \ \left(\frac{\pi}{8}, \boxed{}\right), \ \left(\frac{\pi}{4}, 0\right), \ \left(\boxed{}, \boxed{}\right), \ \left(\frac{\pi}{2}, 0\right).$$

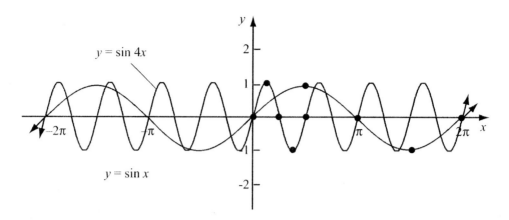

The Constant C: Translating Horizontally $\dfrac{C}{B}$

For each of the functions of the form
$$y = A \sin(Bx - C) + D \quad \text{and} \quad y = A \cos(Bx - C) + D:$$

- If $B = 1$, then the constant C translates the graph horizontally to the right C units if $C > 0$ and to the left $|C|$ units if $C < 0$.

- If $B \neq 1$ and $B > 0$, then $\dfrac{C}{B}$ translates the graph horizontally to the right $\dfrac{C}{B}$ units if $\dfrac{C}{B} > 0$ and to the left $\left|\dfrac{C}{B}\right|$ units if $\dfrac{C}{B} < 0$.

Example 5   Sketch a graph of $y = \sin\left(x - \dfrac{\pi}{2}\right)$.

This function is in the form $y = A \sin(Bx - C) + D$. We have $A = 1$, so the amplitude is $|1|$, or 1. We have $B = 1$, so the period is $\left|\dfrac{2\pi}{1}\right|$, or 2π. The number $D = \boxed{}$, so there will not be a vertical translation. The number C is $\dfrac{\pi}{2} > 0$ and $B = 1$, so there will be a horizontal translation to the $\underset{\text{right / left}}{\underline{}}$ $\dfrac{\pi}{2}$ units. This means that we add $\dfrac{\pi}{2}$ to each x-coordinate on the graph of $y = \sin x$ to get the corresponding x-coordinate on the graph of $y = \sin\left(x - \dfrac{\pi}{2}\right)$.

Consider the following key points on the graph of $y = \sin x$:

$$(0, 0), \quad \left(\frac{\pi}{2}, 1\right), \quad (\pi, 0), \quad \left(\frac{3\pi}{2}, -1\right), \quad (2\pi, 0).$$

These are transformed to

$$\left(\boxed{}, 0\right), \ \left(\pi, \boxed{}\right), \ \left(\boxed{}, 0\right), \ \left(\boxed{}, -1\right), \ \left(\frac{5\pi}{2}, \boxed{}\right).$$

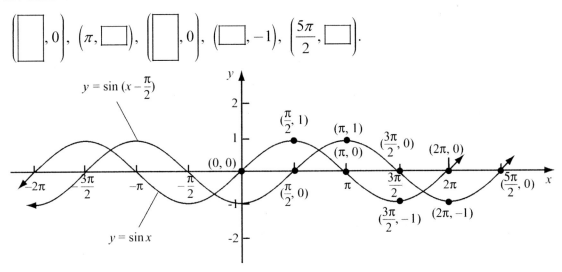

Phase Shift

The **phase shift** of the graphs

$$y = A\sin(Bx - C) + D = A\sin\left[B\left(x - \frac{C}{B}\right)\right] + D$$

and

$$y = A\cos(Bx - C) + D = A\cos\left[B\left(x - \frac{C}{B}\right)\right] + D \ \text{ is } \ \frac{C}{B}.$$

Example 6   Sketch a graph of $y = \cos(2x - \pi)$.

We can write this function in the form

$$y = 1\cdot\cos\left[2\left(x - \frac{\pi}{2}\right)\right] + 0.$$

Amplitude $= |1| = 1$

$$\text{Period} = \left|\frac{2\pi}{\boxed{}}\right| = |\pi| = \boxed{}$$

Translate $\dfrac{\pi}{2}$ units to the $\underline{\hphantom{xxxx}}$. This is the phase shift.
 right / left

The $\boxed{}$ indicates that there is not a vertical translation.

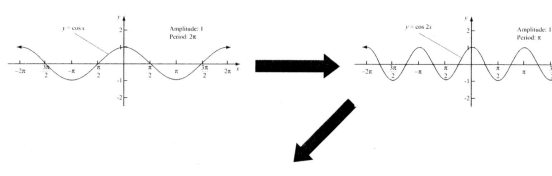

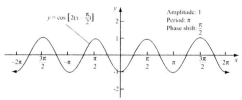

Transformations of the Sine and Cosine Functions

To graph

$$y = A \sin\left(Bx - C\right) + D = A \sin\left[B\left(x - \frac{C}{B}\right)\right] + D$$

and

$$y = A \cos\left(Bx - C\right) + D = A \cos\left[B\left(x - \frac{C}{B}\right)\right] + D,$$

follow the steps listed below in the order in which they are listed.

1. Stretch or shrink the graph horizontally according to B.

    $|B| < 1$ Stretch horizontally

    $|B| > 1$ Shrink horizontally

    $B < 0$ Reflect across the y-axis

    The *period* is $\left|\dfrac{2\pi}{B}\right|$.

2. Stretch or shrink the graph vertically according to A.

    $|A| < 1$ Shrink vertically

    $|A| > 1$ Stretch vertically

    $A < 0$ Reflect across the x-axis

    The *amplitude* is $|A|$.

(continued)

Transformations of the Sine and the Cosine Functions   (continued)

3. Translate the graph horizontally according to C / B.

$\dfrac{C}{B} < 0$ $\left|\dfrac{C}{B}\right|$ units to the left

$\dfrac{C}{B} > 0$ $\dfrac{C}{B}$ units to the right

The *phase shift* is $\dfrac{C}{B}$.

4. Translate the graph vertically according to D.

$D < 0$ $|D|$ units down

$D > 0$ D units up

Example 7   Sketch a graph of $y = 3\sin(2x + \pi/2) + 1$. Find the amplitude, the period, and the phase shift.

Writing this in the form $y = A\sin(Bx - C) + D$, we have $y = 3\sin(2x - (-\pi/2)) + 1$.

Amplitude $= |A| = |3| = \boxed{}$

Period $= \left|\dfrac{2\pi}{B}\right| = \left|\dfrac{2\pi}{\boxed{}}\right| = |\pi| = \pi$

Phase shift $= \dfrac{C}{B} = \dfrac{-\dfrac{\pi}{2}}{2} = -\dfrac{\pi}{\boxed{}}$.   Translate $\dfrac{\pi}{4}$ units to the $\underset{\text{right / left}}{\underline{}}$.

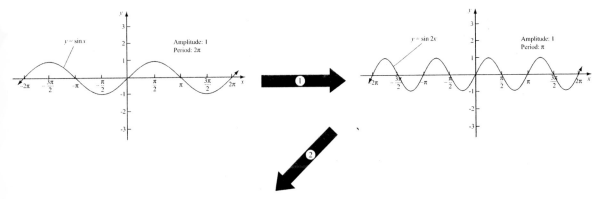

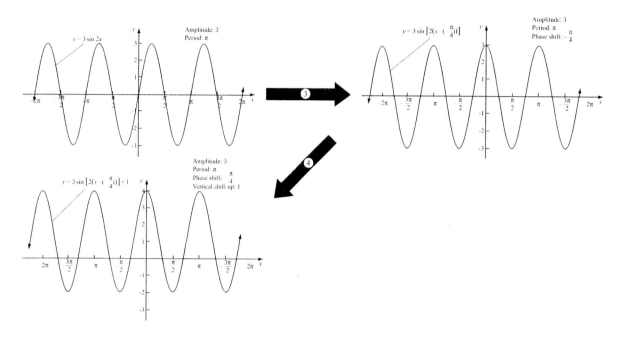

Example 8   Graph $y = 3\cos 2\pi x - 1$. Find the amplitude, the period, and the phase shift.

Consider $y = A\cos(Bx - C) + D$.

$$\text{Amplitude} = |A| = \left|\boxed{}\right| = 3$$

$$\text{Period} = \left|\frac{2\pi}{B}\right| = \left|\frac{2\pi}{\boxed{}}\right| = |1| = 1$$

$$\text{Phase shift} = \frac{C}{B} = \frac{0}{\boxed{}} = 0$$

The value of D, -1, tells us we will translate 1 unit _____ .
 up / down

Graph this on a graphing calculator.

$$y = 3\cos(2\pi x) - 1$$

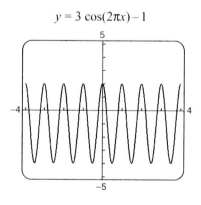

Example 9   Graph: $y = 2\sin x + \sin 2x$.

We graph $y = 2\sin x$ and $y = \sin 2x$ using the same set of axes. Then we add y-coordinates, or ordinates.

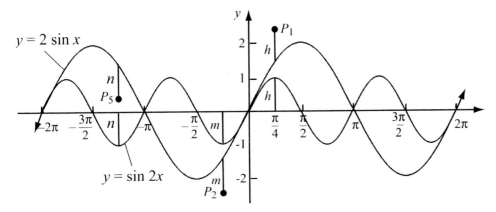

Sketch the graph using the y-values found by adding y-coordinates.

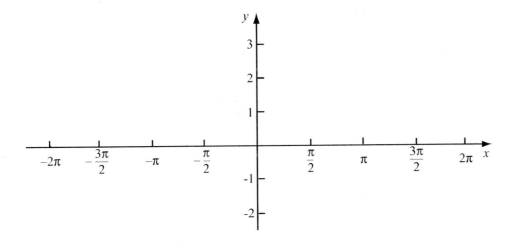

Example 10   Graph $y = 2\cos x - \sin 3x$ and determine its period.

Using a graphing calculator, we graph $Y1 = 2\cos x - \sin 3x$ in the window $[-4\pi, \, 4\pi, \, -4, \, 4]$, $Xscl = \pi / 2$.

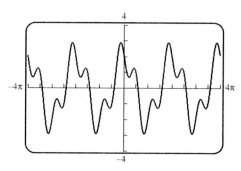

The period appears to be [ ].

Example 11   Sketch a graph of $f(x) = e^{-x/2} \sin x$.

This is an example of damped oscillation. We have the product of two functions, $g(x) = e^{-x/2}$ and $h(x) = \sin x$.

We know that $-1 \leq \sin x \leq$ [ ] and that all of the values of an exponential function are positive. Thus,

$$-e^{-x/2} \leq e^{-x/2} \sin x \leq e^{-x/2}.$$

The graph of $f(x) = e^{-x/2} \sin x$ will be constrained between the graphs of $y = -e^{-x/2}$ and $y = e^{-x/2}$.

Also, the zeros of this function will occur only where $\sin x = 0$, or at $x = k\pi$, k an integer. Use a calculator to find other function values and then draw the graph.

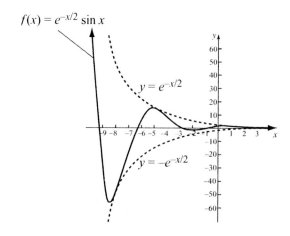

Section 7.1 Identities: Pythagorean and Sum and Difference

An **identity** is an equation that is true for all *possible* replacements of the variables.

Basic Identities

$$\sin x = \frac{1}{\csc x}, \qquad \csc x = \frac{1}{\sin x}, \qquad \sin(-x) = -\sin x,$$

$$\cos(-x) = \cos x,$$

$$\cos x = \frac{1}{\sec x}, \qquad \sec x = \frac{1}{\cos x}, \qquad \tan(-x) = -\tan x,$$

$$\tan x = \frac{1}{\cot x}, \qquad \cot x = \frac{1}{\tan x}, \qquad \tan x = \frac{\sin x}{\cos x},$$

$$\cot x = \frac{\cos x}{\sin x}$$

Pythagorean Identities	Equivalent Forms
$\sin^2 x + \cos^2 x = 1$	$\sin^2 x = 1 - \cos^2 x$ $\cos^2 x = 1 - \sin^2 x$
$1 + \cot^2 x = \csc^2 x$	$1 = \csc^2 x - \cot^2 x$ $\cot^2 x = \csc^2 x - 1$
$1 + \tan^2 x = \sec^2 x$	$1 = \sec^2 x - \tan^2 x$ $\tan^2 x = \sec^2 x - 1$

Example 1 Multiply and simplify: $\cos x (\tan x - \sec x)$.

$$\cos x (\tan x - \sec x)$$

$$= \cos x \cdot \tan x - \boxed{} \cdot \sec x$$

$$= \cos x \cdot \frac{\sin x}{\cos x} - \cos x \cdot \frac{1}{\boxed{}}$$

$$= \sin x - \boxed{}$$

Example 2 Factor and simplify: $\sin^2 x \cos^2 x + \cos^4 x$.

$$\sin^2 x \cos^2 x + \cos^4 x$$
$$= \sin^2 x \cdot \cos^2 x + \cos^2 x \cdot \cos^2 x$$
$$= \cos^2 x \left(\sin^2 x + \boxed{}\right)$$
$$= \cos^2 x \cdot 1 \qquad \left(\sin^2 x + \cos^2 x = 1\right)$$
$$= \boxed{}$$

Example 3 Simplify each of the following trigonometric expressions.

a) $\dfrac{\cot(-\theta)}{\csc(-\theta)}$

b) $\dfrac{2\sin^2 t + \sin t - 3}{1 - \cos^2 t - \sin t}$

a) $\dfrac{\cot(-\theta)}{\csc(-\theta)} = \dfrac{\dfrac{\cos(-\theta)}{\sin(-\theta)}}{\dfrac{1}{\boxed{}}}$

$$= \dfrac{\cos(-\theta)}{\sin(-\theta)} \cdot \dfrac{\boxed{}}{1}$$
$$= \cos(-\theta)$$
$$= \cos\theta$$

b) $\dfrac{2\sin^2 + \sin t - 3}{1 - \cos^2 t - \sin t}$

$$= \dfrac{2\sin^2 t + \sin t - 3}{\sin^2 t - \sin t} \qquad \left(1 - \cos^2 t = \sin^2 t\right)$$
$$= \dfrac{\left(2\sin t + 3\right)\left(\sin t - \boxed{}\right)}{\sin t \left(\sin t - \boxed{}\right)}$$
$$= \dfrac{2\sin t + 3}{\sin t}$$
$$= 2 + \dfrac{3}{\boxed{}}, \text{ or } 2 + 3\boxed{}$$

Example 4 Add and simplify: $\dfrac{\cos x}{1+\sin x} + \tan x$.

$$\dfrac{\cos x}{1+\sin x} + \tan x = \dfrac{\cos x}{1+\sin x} + \dfrac{\sin x}{\boxed{}}$$

$$= \dfrac{\cos x}{1+\sin x} \cdot \dfrac{\cos x}{\cos x} + \dfrac{\sin x}{\cos x} \cdot \dfrac{1+\sin x}{1+\boxed{}}$$

$$= \dfrac{\cos^2 x + \sin x + \sin^2 x}{\cos x (1+\sin x)}$$

$$= \dfrac{1+\sin x}{\boxed{} (1+\sin x)}$$

$$= \dfrac{1}{\cos x}, \text{ or } \boxed{}$$

Example 5 Multiply and simplify: $\sqrt{\sin^3 x \cos x} \cdot \sqrt{\cos x}$.

$$\sqrt{\sin^3 x \cos x} \cdot \sqrt{\cos x}$$

$$= \sqrt{\sin^3 x \cos x \cdot \boxed{}}$$

$$= \sqrt{\sin^3 x \cdot \cos^2 x}$$

$$= \sqrt{\sin^2 x \cdot \boxed{} \cdot \cos^2 x}$$

$$= \sqrt{\sin^2 x} \cdot \sqrt{\cos^2 x} \cdot \sqrt{\boxed{}}$$

$$= \sin x \cos x \cdot \sqrt{\sin x}$$

Example 6 Rationalize the denominator: $\sqrt{\dfrac{2}{\tan x}}$.

$$\sqrt{\dfrac{2}{\tan x}} = \sqrt{\dfrac{2}{\tan x} \cdot \dfrac{\tan x}{\boxed{}}}$$

$$= \sqrt{\dfrac{2\tan x}{\tan^2 x}}$$

$$= \dfrac{\sqrt{2\tan x}}{\sqrt{\boxed{}}} = \dfrac{\sqrt{2\tan x}}{\tan x}$$

Example 7   Express $\sqrt{9+x^2}$ as a trigonometric function of θ without using radicals by letting $x=3\tan\theta$. Assume that $0<\theta<\pi/2$. Then find $\sin\theta$ and $\cos\theta$.

$$x=3\tan\theta$$

$$\sqrt{9+x^2}=\sqrt{9+\left(3\tan\theta\right)^2}$$

$$=\sqrt{\boxed{}+9\tan^2\theta}$$

$$=\sqrt{\boxed{}\left(1+\tan^2\theta\right)}$$

$$=\sqrt{9\sec^2\theta}\qquad\left(1+\tan^2\theta=\sec^2\theta\right)$$

$$=\boxed{}\left|\sec\theta\right|=3\sec\theta$$

We have $\sqrt{9+x^2}=3\sec\theta$, so $\sec\theta=\dfrac{\sqrt{9+x^2}}{\boxed{}}$.

We draw a right triangle containing θ.

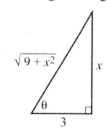

$$\sin\theta=\frac{\boxed{}}{\sqrt{9+x^2}}$$

$$\cos\theta=\frac{\boxed{}}{\sqrt{9+x^2}}$$

Sum and Difference Identities

$$\sin\left(u\pm v\right)=\sin u\cos v\pm\cos u\sin v,$$

$$\cos\left(u\pm v\right)=\cos u\cos v\mp\sin u\sin v,$$

$$\tan\left(u\pm v\right)=\frac{\tan u\pm\tan v}{1\mp\tan u\tan v}$$

Example 8   Find $\cos(5\pi/12)$ exactly.

$$\frac{5\pi}{12} = \frac{9\pi}{12} - \frac{\boxed{}\pi}{12} = \frac{3\pi}{4} - \frac{\pi}{\boxed{}}$$

$$\cos(u-v) = \cos u \cos v + \sin u \sin v$$

$$\cos\frac{5\pi}{12} = \cos\left(\frac{3\pi}{4} - \frac{\pi}{3}\right) \qquad \left(u = \frac{3\pi}{4},\ v = \frac{\pi}{3}\right)$$

$$= \cos\frac{3\pi}{4}\cos\frac{\pi}{3} + \sin\frac{3\pi}{4}\sin\boxed{}$$

$$= -\frac{\sqrt{2}}{2}\cdot\frac{1}{2} + \frac{\sqrt{2}}{2}\cdot\frac{\sqrt{3}}{\boxed{}}$$

$$= -\frac{\sqrt{2}}{4} + \frac{\sqrt{6}}{4} = \frac{-\sqrt{2}+\sqrt{6}}{\boxed{}} = \frac{\sqrt{6}-\sqrt{2}}{4}$$

Example 9   Find $\sin 105°$ exactly.

$$105° = 45° + \boxed{}°$$

$$\sin 105° = \sin(45° + 60°)$$

$$\sin(u+v) = \sin u \cos v + \cos u \sin v$$

Let $u = 45°$ and $v = 60°$.

$$\sin 105° = \sin 45°\cos\boxed{}° + \cos\boxed{}°\sin 60°$$

$$= \frac{\sqrt{2}}{2}\cdot\frac{1}{2} + \frac{\sqrt{2}}{2}\cdot\frac{\sqrt{3}}{\boxed{}}$$

$$= \frac{\sqrt{2}}{4} + \frac{\sqrt{6}}{\boxed{}}$$

$$= \frac{\sqrt{2}+\sqrt{6}}{4}$$

Example 10   Find $\tan 15°$ exactly.

$$15° = 45° - \boxed{}°$$

$$\tan(u-v) = \frac{\tan u - \tan v}{1 + \tan u \tan v}$$

$$\tan 15° = \tan(45° - 30°) \qquad\qquad (u = 45°,\ v = 30°)$$

$$= \frac{\tan 45° - \tan \boxed{}°}{1 + \tan \boxed{}° \tan 30°} = \frac{1 - \dfrac{\sqrt{3}}{3}}{1 + 1 \cdot \dfrac{\sqrt{3}}{3}}$$

$$= \frac{\dfrac{3-\sqrt{3}}{3}}{\dfrac{3+\sqrt{3}}{\boxed{}}} = \frac{3-\sqrt{3}}{3} \cdot \frac{3}{\boxed{}+\sqrt{3}} = \frac{3-\sqrt{3}}{3+\sqrt{3}}$$

Example 11   Assume that $\sin\alpha = \dfrac{2}{3}$ and $\sin\beta = \dfrac{1}{3}$ and that α and β are between 0 and $\pi/2$. Then evaluate $\sin(\alpha+\beta)$.

$$\sin(\alpha+\beta) = \sin\alpha\cos\beta + \cos\alpha\sin\beta$$

$$= \frac{2}{3}\cdot\cos\beta + \cos\alpha\cdot\frac{1}{3}$$

Using reference triangles and the Pythagorean theorem, we determine the values of $\cos\beta$ and $\cos\alpha$.

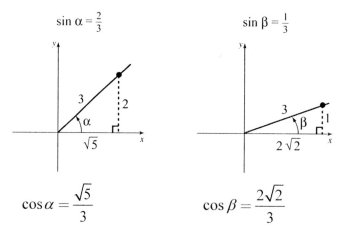

$$\sin\alpha = \tfrac{2}{3} \qquad\qquad\qquad \sin\beta = \tfrac{1}{3}$$

$$\cos\alpha = \frac{\sqrt{5}}{3} \qquad\qquad\qquad \cos\beta = \frac{2\sqrt{2}}{3}$$

Now we have

$$\sin(\alpha + \beta) = \frac{2}{3} \cdot \frac{2\sqrt{2}}{\Box} + \frac{\sqrt{5}}{3} \cdot \frac{1}{3}$$

$$= \frac{4\sqrt{2}}{9} + \frac{\sqrt{5}}{9} = \frac{4\sqrt{2} + \sqrt{5}}{\Box}.$$

Example 12 Assume that $\cos\alpha = -\dfrac{4}{5}$ with α between π and $3\pi/2$ and that

$\cos\beta = -\dfrac{2}{5}$ with β between $\pi/2$ and π. Then evaluate $\cos(\alpha - \beta)$.

$$\cos(\alpha - \beta) = \cos\alpha \cos\beta + \sin\alpha \sin\beta$$

We know the values of $\cos\alpha$ and $\cos\beta$. Using reference triangles and the Pythagorean theorem, we determine the values of $\sin\alpha$ and $\sin\beta$.

$$\cos\alpha = -\frac{4}{5} \qquad\qquad \cos\beta = -\frac{2}{5}$$

$$\sin\alpha = -\frac{3}{5} \qquad\qquad \sin\beta = \frac{\sqrt{21}}{5}$$

We have

$$\cos(\alpha - \beta) = \left(-\frac{4}{5}\right)\left(-\frac{2}{5}\right) + \left(-\frac{3}{5}\right)\left(\frac{\sqrt{21}}{\Box}\right)$$

$$= \frac{\Box}{25} + \left(\frac{-3\sqrt{21}}{\Box}\right)$$

$$= \frac{8 - \Box\sqrt{21}}{25}.$$

Section 7.2   Identities: Cofunction, Double-Angle, and Half-Angle

Cofunction Identities

$$\sin\left(\frac{\pi}{2} - x\right) = \cos x, \qquad \cos\left(\frac{\pi}{2} - x\right) = \sin x, \qquad \sin\left(x + \frac{\pi}{2}\right) = \cos x,$$

$$\tan\left(\frac{\pi}{2} - x\right) = \cot x, \qquad \cot\left(\frac{\pi}{2} - x\right) = \tan x, \qquad \sin\left(x - \frac{\pi}{2}\right) = -\cos x,$$

$$\sec\left(\frac{\pi}{2} - x\right) = \csc x, \qquad \csc\left(\frac{\pi}{2} - x\right) = \sec x, \qquad \cos\left(x + \frac{\pi}{2}\right) = -\sin x,$$

$$\cos\left(x - \frac{\pi}{2}\right) = \sin x$$

Example 1   Prove the identity $\sin(x + \pi/2) = \cos x$.

$$\sin(u + v) = \sin u \cos v + \cos u \sin v$$

$$\sin(x + \pi/2) = \sin x \cos\frac{\pi}{2} + \cos x \sin\frac{\pi}{2} \qquad (u = x,\ v = \pi/2)$$

$$= \sin x \cdot \boxed{} + \cos x \cdot \boxed{}$$

$$= 0 + \cos x = \cos x$$

Example 2   Find an identity for each of the following.

a) $\tan\left(x + \dfrac{\pi}{2}\right)$

b) $\sec(x - 90°)$

a) $\tan\left(x + \dfrac{\pi}{2}\right) = \dfrac{\sin\left(x + \dfrac{\pi}{2}\right)}{\cos\left(x + \dfrac{\pi}{2}\right)}$

$$= \frac{\cos\boxed{}}{-\sin x}$$

$$= -\boxed{}x$$

We have the identity

$$\tan\left(x + \frac{\pi}{2}\right) = -\cot x.$$

b) $\sec\left(x-90°\right)=\dfrac{1}{\boxed{}\left(x-90°\right)}$

$=\dfrac{1}{\sin x}$

$=\boxed{}x$

We have the identity

$\sec\left(x-90°\right)=\csc x.$

Double-Angle Identities

$\sin 2x = 2\sin x\cos x,\qquad\qquad \cos 2x = \cos^{2}x-\sin^{2}x,$

$\tan 2x = \dfrac{2\tan x}{1-\tan^{2}x}\qquad\qquad\quad =1-2\sin^{2}x$

$\qquad\qquad\qquad\qquad\qquad\quad =2\cos^{2}x-1$

From the last two cosine double-angle identities above, we can obtain three addition identities:

$$\sin^{2}x=\dfrac{1-\cos 2x}{2},\quad \cos^{2}x=\dfrac{1+\cos 2x}{2},\quad \tan^{2}x=\dfrac{1-\cos 2x}{1+\cos 2x}.$$

Example 3   Given that $\tan\theta=-\dfrac{3}{4}$ and θ is in quadrant II, find each of the following.

a) $\sin 2\theta$

b) $\cos 2\theta$

c) $\tan 2\theta$

d)   The quadrant in which 2θ lies

We first draw a reference triangle in order to determine the values of $\sin\theta$ and $\cos\theta$.

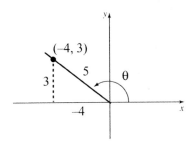

(-4, 3)

5

3

-4

θ

$\sin\theta=\dfrac{3}{5},\quad \cos\theta=\dfrac{-4}{5}=-\dfrac{4}{5}$

a) $\sin 2\theta = 2\sin\theta\cos\theta$

$=2\cdot\boxed{}\cdot\left(-\dfrac{4}{5}\right)=-\dfrac{\boxed{}}{25}$

b) $\cos 2\theta = \cos^2 \theta - \sin^2 \theta$

$$= \left(-\frac{4}{5}\right)^2 - \left(\boxed{}\right)^2$$

$$= \frac{16}{\boxed{}} - \frac{9}{25} = \frac{\boxed{}}{25}$$

c)   We can use the values we found for $\sin 2\theta$ and $\cos 2\theta$ to find $\tan 2\theta$.

$$\tan 2\theta = \frac{\sin 2\theta}{\cos 2\theta} = \frac{-\dfrac{24}{25}}{\dfrac{7}{25}} = -\frac{\boxed{}}{7}$$

We can also use a double-angle identity.

$$\tan 2\theta = \frac{2\tan \theta}{1 - \tan^2 \theta}$$

$$= \frac{2\left(-\dfrac{3}{4}\right)}{1 - \left(\boxed{}\right)^2}$$

$$= \frac{-\dfrac{3}{2}}{1 - \dfrac{9}{16}} = \frac{-\dfrac{3}{2}}{\dfrac{7}{16}}$$

$$= -\frac{3}{2} \cdot \frac{16}{7} = -\frac{\boxed{}}{7}$$

d)   We found that $\sin 2\theta$ is _____ , and $\cos 2\theta$ is _____ , so
positive / negative   positive / negative

2θ is in quadrant $\boxed{}$.

Example 4   Find an equivalent expression for each of the following.

a) $\sin 3\theta$ in terms of function values of θ

b) $\cos^3 x$ in terms of function values of x or $2x$, raised only to the first power.

a) $\sin 3\theta = \sin \left(2\theta + \boxed{}\right)$

$\qquad = \sin 2\theta \cos \theta + \cos \boxed{} \sin \theta$

$\qquad = 2\sin \theta \cos \theta \cos \theta + \left(2\cos^2 \theta - \boxed{}\right)\sin \theta$

$\qquad = 2\sin \theta \cos^2 \theta + 2\cos^2 \theta \sin \theta - \boxed{}$

$\qquad = \boxed{} \sin \theta \cos^2 \theta - \sin \theta$

b) $\cos^3 x = \cos^2 x \cos x$

$\qquad = \dfrac{\cos 2x + 1}{2} \cdot \boxed{}$

$\qquad = \dfrac{\cos x \cos 2x + \cos x}{\boxed{}}$

Half-Angle Identities

$$\sin \frac{x}{2} = \pm\sqrt{\frac{1 - \cos x}{2}},$$

$$\cos \frac{x}{2} = \pm\sqrt{\frac{1 + \cos x}{2}},$$

$$\tan \frac{x}{2} = \pm\sqrt{\frac{1 - \cos x}{1 + \cos x}}$$

$$= \frac{\sin x}{1 + \cos x} = \frac{1 - \cos x}{\sin x}$$

The use of $+$ or $-$ depends on the quadrant in which the angle $x/2$ lies.

Example 5 Find $\tan\left(\pi/8\right)$ exactly.

$$\frac{\pi}{8} = \frac{1}{2}\cdot\frac{\pi}{4} = \frac{\frac{\pi}{4}}{2}$$

$$\tan\frac{x}{2} = \frac{\sin x}{1+\cos x}$$

$$\tan\frac{\frac{\pi}{4}}{2} = \frac{\sin\frac{\pi}{4}}{1+\cos\frac{\pi}{4}} \qquad \left(x = \frac{\pi}{4}\right)$$

$$= \frac{\frac{\sqrt{2}}{\square}}{1+\frac{\square}{2}} = \frac{\frac{\sqrt{2}}{2}}{\frac{\square+\sqrt{2}}{2}}$$

$$= \frac{\sqrt{2}}{2}\cdot\frac{2}{2+\sqrt{2}} = \frac{\square}{2+\sqrt{2}}$$

$$= \frac{\sqrt{2}}{2+\sqrt{2}}\cdot\frac{2-\sqrt{2}}{2-\sqrt{2}} \qquad \text{Rationalizing the denominator}$$

$$= \frac{2\sqrt{2}-\square}{\square-2} = \frac{2\left(\sqrt{2}-1\right)}{2\cdot1}$$

$$= \sqrt{2}-1$$

Example 6 Simplify each of the following.

a) $\dfrac{\sin x\cos x}{\dfrac{1}{2}\cos 2x}$

b) $2\sin^2\dfrac{x}{2}+\cos x$

a) $\dfrac{\sin x\cos x}{\dfrac{1}{2}\cos 2x} = \dfrac{2}{2}\cdot\dfrac{\sin x\cos x}{\dfrac{1}{2}\cos 2x}$

$$= \frac{\square\,\sin x\cos x}{\cos 2x}$$

$$= \frac{\sin\square}{\cos 2x}$$

$$= \boxed{}2x$$

Section 7.2 Identities: Cofunction, Double-Angle, and Half-Angle   213

b) $2\sin^2\dfrac{x}{2}+\cos x$

Consider the identity

$$\sin\frac{x}{2}=\pm\sqrt{\frac{1-\cos x}{2}}.$$

Squaring both sides, we have

$$\sin^2\frac{x}{2}=\frac{1-\cos x}{2}.$$

Then we have

$$2\sin^2\frac{x}{2}+\cos x=2\cdot\frac{1-\cos x}{2}+\boxed{}$$

$$=\boxed{}-\cos x+\cos x$$

$$=\boxed{}.$$

Copyright © 2017 Pearson Education, Inc.

Section 7.3 Proving Trigonometric Identities

> **Method 1:** Start with either side of the equation and obtain the other side.
>
> **Method 2:** Work with each side separately until you obtain the same expression.

Example 1   Prove the identity $1 + \sin 2\theta = (\sin\theta + \cos\theta)^2$.

We will use method 1, beginning with the right side.

$$(\sin\theta + \cos\theta)^2$$
$$= \sin^2\theta + \boxed{}\sin\theta\cos\theta + \cos^2\theta$$
$$= 1 + 2\sin\theta\cos\theta \qquad\qquad (\sin^2\theta + \cos^2\theta = 1)$$
$$= 1 + \sin\boxed{} \qquad\qquad (\sin 2\theta = 2\sin\theta\cos\theta)$$

We could also begin with the left side.

$$1 + \sin 2\theta = 1 + \boxed{}\sin\theta\cos\theta = \sin^2\theta + 2\sin\theta\cos\theta + \cos^2\theta = \left(\boxed{} + \cos\theta\right)^2$$

Example 2   Prove the identity $\sin^2 x \tan^2 x = \tan^2 x - \sin^2 x$.

We will work with each side separately using method 2. We start with the left side.

$$\sin^2 x \tan^2 x = \sin^2 x \cdot \frac{\sin^2 x}{\boxed{}} = \frac{\sin^4 x}{\cos^2 x}$$

Now we work with the right side.

$$\tan^2 x - \sin^2 x = \frac{\sin^2 x}{\cos^2 x} - \boxed{}$$
$$= \frac{\sin^2 x - \sin^2 x \cos^2 x}{\cos^2 x}$$
$$= \frac{\sin^2 x\left(\boxed{} - \cos^2 x\right)}{\cos^2 x}$$
$$= \frac{\sin^2 x \cdot \sin^2 x}{\cos^2 x} \qquad\qquad \left(1 - \cos^2 x = \sin^2 x\right)$$
$$= \frac{\sin^4 x}{\cos^2 x}$$

This is the same expression we obtained when we worked with the left side, so we have proved the identity.

Example 3   Prove the identity $\dfrac{\sin 2x}{\sin x} - \dfrac{\cos 2x}{\cos x} = \sec x$.

We will use method 1, starting with the left side.

$$\dfrac{\sin 2x}{\sin x} - \dfrac{\cos 2x}{\cos x}$$

$$= \dfrac{\boxed{}\sin x \cos x}{\sin x} - \dfrac{\cos^2 x - \sin^2 x}{\boxed{}}$$

$$= 2\cos x - \dfrac{\cos^2 x - \sin^2 x}{\cos x}$$

$$= \dfrac{2\cos^2 x - \cos^2 x + \sin^2 x}{\boxed{}}$$

$$= \dfrac{\cos^2 x + \sin^2 x}{\cos x}$$

$$= \dfrac{\boxed{}}{\cos x}$$

$$= \sec x$$

Example 4   Prove the identity $\dfrac{\sec t - 1}{t \sec t} = \dfrac{1 - \cos t}{t}$.

We will use method 1, starting with the left side.

$$\dfrac{\sec t - 1}{t \sec t} = \dfrac{\dfrac{1}{\cos t} - 1}{t \cdot \dfrac{1}{\boxed{}}} = \dfrac{\dfrac{1}{\cos t} - 1}{\dfrac{t}{\cos t}}$$

$$= \left(\dfrac{1}{\cos t} - 1\right) \cdot \dfrac{\cos t}{\boxed{}}$$

$$= \dfrac{1}{t} - \dfrac{\cos t}{t} = \dfrac{\boxed{} - \cos t}{t}$$

Example 5   Prove the identity $\cot \phi + \csc \phi = \dfrac{\sin \phi}{1 - \cos \phi}$.

We will use method 2, beginning with the left side.

$$\cot \phi + \csc \phi = \dfrac{\cos \phi}{\boxed{}} + \dfrac{1}{\sin \phi}$$

$$= \dfrac{\cos \phi + \boxed{}}{\sin \phi}$$

Now we work with the right side.

$$\frac{\sin\phi}{1-\cos\phi} = \frac{\sin\phi}{1-\cos\phi} \cdot \frac{1+\cos\phi}{1+\cos\phi}$$

$$= \frac{\sin\phi\left(\boxed{}+\cos\phi\right)}{1-\cos^2\phi}$$

$$= \frac{\sin\phi\left(1+\cos\phi\right)}{\sin^2\phi} \qquad \left(1-\cos^2\phi = \sin^2\phi\right)$$

$$= \frac{1+\cos\phi}{\boxed{}}$$

We obtained the same expression from each side, so we have proved the identity.

Product-to-Sum Identities

$$\sin x \cdot \sin y = \frac{1}{2}\left[\cos(x-y) - \cos(x+y)\right]$$

$$\cos x \cdot \cos y = \frac{1}{2}\left[\cos(x-y) + \cos(x+y)\right]$$

$$\sin x \cdot \cos y = \frac{1}{2}\left[\sin(x+y) + \sin(x-y)\right]$$

$$\cos x \cdot \sin y = \frac{1}{2}\left[\sin(x+y) - \sin(x-y)\right]$$

Example 6 Find an identity for $2\sin 3\theta \cos 7\theta$.

We will use the identity

$$\sin x \cdot \cos y = \frac{1}{2}\left[\sin(x+y) + \sin(x-y)\right]$$

with $x = 3\theta$ and $y = 7\theta$.

$$2\sin 3\theta \cos 7\theta$$

$$= 2 \cdot \frac{1}{2}\left[\sin(3\theta + 7\theta) + \sin(3\theta - 7\theta)\right]$$

$$= \sin\boxed{} + \sin(-4\theta)$$

$$= \sin 10\theta - \sin\boxed{} \qquad \left(\sin(-x) = -\sin x\right)$$

Sum-to-Product Identities

$$\sin x + \sin y = 2\sin\frac{x+y}{2}\cos\frac{x-y}{2}$$

$$\sin x - \sin y = 2\cos\frac{x+y}{2}\sin\frac{x-y}{2}$$

$$\cos y + \cos x = 2\cos\frac{x+y}{2}\cos\frac{x-y}{2}$$

$$\cos y - \cos x = 2\sin\frac{x+y}{2}\sin\frac{x-y}{2}$$

Example 7   Find an identity for $\cos\theta + \cos 5\theta$.

We will use the identity

$$\cos y + \cos x = 2\cos\frac{x+y}{2}\cos\frac{x-y}{2}$$

with $x = 5\theta$ and $y = \theta$.

$$\cos\theta + \cos 5\theta$$

$$= 2\cos\frac{\boxed{}+\theta}{2}\cos\frac{5\theta-\boxed{}}{2}$$

$$= 2\cos\frac{\boxed{}}{2}\cos\frac{4\theta}{2}$$

$$= 2\cos 3\theta\cos\boxed{}$$

Section 7.4 Inverses of the Trigonometric Functions

Inverse Trigonometric Functions

FUNCTION	DOMAIN	RANGE
$y = \sin^{-1} x$ $= \arcsin x,$ where $x = \sin y$	$[-1, 1]$	$[-\pi/2, \pi/2]$
$y = \cos^{-1} x$ $= \arccos x,$ where $x = \cos y$	$[-1, 1]$	$[0, \pi]$
$y = \tan^{-1} x$ $= \arctan x,$ where $x = \tan y$	$(-\infty, \infty)$	$(-\pi/2, \pi/2)$
$y = \csc^{-1} x$	$(-\infty, -1] \cup [1, \infty)$	$[-\pi/2, 0) \cup (0, \pi/2]$
$y = \sec^{-1} x$	$(-\infty, -1] \cup [1, \infty)$	$[0, \pi/2) \cup (\pi/2, \pi]$
$y = \cot^{-1} x$	$(-\infty, \infty)$	$(0, \pi)$

Example 1   Find each of the following function values.

a) $\sin^{-1} \dfrac{\sqrt{2}}{2}$ b) $\cos^{-1}\left(-\dfrac{1}{2}\right)$ c) $\tan^{-1}\left(-\dfrac{\sqrt{3}}{3}\right)$

d) $\cot^{-1} 0$ e) $\sec^{-1}\left(-\sqrt{2}\right)$

a)   The range is $\left[-\dfrac{\pi}{2}, \dfrac{\pi}{2}\right]$. The only number in $\left[-\dfrac{\pi}{2}, \dfrac{\pi}{2}\right]$ with a sine of $\dfrac{\sqrt{2}}{2}$ is $\dfrac{\pi}{4}$, so

$\sin^{-1} \dfrac{\sqrt{2}}{2} = \boxed{}$.

b)   The range is $[0, \pi]$. The only number in $[0, \pi]$ with a cosine of $-\dfrac{1}{2}$ is $\dfrac{2\pi}{3}$, so

$\cos^{-1}\left(-\dfrac{1}{2}\right) = \boxed{}$.

c)   The range is $\left(-\dfrac{\pi}{2}, \dfrac{\pi}{2}\right)$. The only number in $\left(-\dfrac{\pi}{2}, \dfrac{\pi}{2}\right)$ with a tangent of $-\dfrac{\sqrt{3}}{3}$ is $-\dfrac{\pi}{6}$,

so $\tan^{-1}\left(-\dfrac{\sqrt{3}}{3}\right) = \boxed{}$.

d)   The range is $(0, \pi)$. The only number in $(0, \pi)$ with a cotangent of 0 is $\dfrac{\pi}{2}$, so

$\cot^{-1} 0 = \boxed{}$.

e)   The range is $\left[0, \dfrac{\pi}{2}\right) \cup \left(\dfrac{\pi}{2}, \pi\right]$. The only number in the restricted range with a secant of

$-\sqrt{2}$ is $\dfrac{3\pi}{4}$, so $\sec^{-1}\left(-\sqrt{2}\right) = \boxed{}$.

Example 2   Approximate each of the following function values in both radians and degrees. Round radian measure to four decimal places and degree measure to the nearest tenth of a degree.

a) $\cos^{-1}(-0.2689)$

b) $\tan^{-1}(-0.2623)$

c) $\sin^{-1} 0.20345$

d) $\cos^{-1} 1.318$

e) $\csc^{-1} 8.205$

FUNCTION VALUE	MODE	READOUT	ROUNDED
a) $\cos^{-1}(-0.2689)$	Radian	1.843047111	1.8430
	Degree	105.5988209	$\boxed{}^{\circ}$
b) $\tan^{-1}(-0.2623)$	Radian	$-.2565212141$	-0.2565
	Degree	-14.69758292	-14.7°
c) $\sin^{-1} 0.20345$	Radian	.2048803359	$\boxed{}$
	Degree	11.73877855	11.7°
d) $\cos^{-1} 1.318$	Radian	ERR:DOMAIN	
	Degree	ERR:DOMAIN	

The value 1.318 is not in $[-1, 1]$, the domain of the inverse cosine function.

e) $\csc^{-1} 8.205 =$	Radian	.1221806653	$\boxed{}$
$\sin^{-1}\left(1/\boxed{}\right)$	Degree	7.000436462	$\boxed{}^{\circ}$

Composition of Trigonometric Functions

$\sin\left(\sin^{-1} x\right) = x,$ for all x in the domain of $\sin^{-1}$.

$\cos\left(\cos^{-1} x\right) = x,$ for all x in the domain of $\cos^{-1}$.

$\tan\left(\tan^{-1} x\right) = x,$ for all x in the domain of $\tan^{-1}$.

Example 3   Simplify each of the following.

a) $\cos\left(\cos^{-1}\dfrac{\sqrt{3}}{2}\right)$

b) $\sin\left(\sin^{-1}1.8\right)$

a)   Domain of $\cos^{-1}$: $\left[-1, \boxed{}\right]$

$-1 \le \dfrac{\sqrt{3}}{2} \le 1$, so $\cos\left(\cos^{-1}\dfrac{\sqrt{3}}{2}\right) = \boxed{}$.

b)   Domain of $\sin^{-1}$: $\left[\boxed{}, 1\right]$

$1.8 > 1$, so 1.8 $\underline{\hspace{2cm}}$ in the domain of $\sin^{-1}$.
$\qquad\qquad\qquad\qquad$ is / is not

Thus, $\sin\left(\sin^{-1}1.8\right)$ does not exist.

Special Cases

$\sin^{-1}(\sin x) = x$,   for all x in the range of $\sin^{-1}$.

$\cos^{-1}(\cos x) = x$,   for all x in the range of $\cos^{-1}$.

$\tan^{-1}(\tan x) = x$,   for all x in the range of $\tan^{-1}$.

Example 4   Simplify each of the following.

a) $\tan^{-1}\left(\tan\dfrac{\pi}{6}\right)$

b) $\sin^{-1}\left(\sin\dfrac{3\pi}{4}\right)$

a)   Range of $\tan^{-1}$: $\left(-\dfrac{\pi}{2}, \boxed{}\right)$

$-\dfrac{\pi}{2} < \dfrac{\pi}{6} < \dfrac{\pi}{2}$, so $\tan^{-1}\left(\tan\dfrac{\pi}{6}\right) = \boxed{}$.

b)   Range of $\sin^{-1}$: $\left[-\dfrac{\pi}{2}, \dfrac{\pi}{2}\right]$

$\dfrac{3\pi}{4} > \dfrac{\pi}{2}$, so $\dfrac{3\pi}{4}$ is not in the range of $\sin^{-1}$. We cannot use the identity

$\sin^{-1}(\sin x) = x$, but we can evaluate the expression.

$\sin^{-1}\left(\sin\dfrac{3\pi}{4}\right) = \sin^{-1}\left(\dfrac{\sqrt{2}}{2}\right) = \boxed{}$.

Example 5   Simplify each of the following.

a) $\sin\left[\tan^{-1}(-1)\right]$ b) $\cos^{-1}\left(\sin\dfrac{\pi}{2}\right)$

a)   Let $\tan^{-1}(-1)=\theta$. Find θ such that $\tan\theta=-1$ and $-\dfrac{\pi}{2}<\theta<\dfrac{\pi}{2}$. We have

$\theta=\boxed{}$. Thus, $\sin\left[\tan^{-1}(-1)\right]=\sin\left(-\dfrac{\pi}{4}\right)=\boxed{}$.

b) $\cos^{-1}\left(\sin\dfrac{\pi}{2}\right)=\cos^{-1}\boxed{}=\boxed{}$

Example 6   Find $\sin\left(\cot^{-1}\dfrac{x}{2}\right)$.

Since $\cot^{-1}$ is defined in $(0,\pi)$, we consider quadrants I and II. We make a drawing where $\cot\theta=\dfrac{x}{2}$.

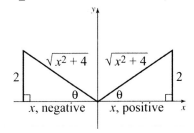

We read the sine ratio from the triangles.

$$\sin\left(\cot^{-1}\dfrac{x}{2}\right)=\dfrac{\boxed{}}{\sqrt{x^2+4}}$$

Example 7   Evaluate: $\sin\left(\sin^{-1}\dfrac{1}{2}+\cos^{-1}\dfrac{5}{13}\right)$.

We have the sine of the sum of two angles. We use the identity

$$\sin(u+v)=\sin u\cos v\ |\boxed{}\sin v.$$

We have $u=\sin^{-1}\dfrac{1}{2}$ and $v=\cos^{-1}\dfrac{5}{13}$.

$$\sin\left(\sin^{-1}\dfrac{1}{2}+\cos^{-1}\dfrac{5}{13}\right)$$
$$=\sin\left(\sin^{-1}\dfrac{1}{2}\right)\cos\left(\cos^{-1}\dfrac{5}{13}\right)+\cos\left(\sin^{-1}\dfrac{1}{2}\right)\sin\left(\cos^{-1}\dfrac{5}{13}\right)$$

We know that $\sin\left(\sin - \dfrac{1}{2}\right) = \boxed{}$ and $\cos\left(\cos^{-1}\dfrac{5}{13}\right) = \boxed{}$. Now consider $\cos\left(\sin^{-1}\dfrac{1}{2}\right)$.

$$\cos\left(\sin^{-1}\dfrac{1}{2}\right) = \cos\dfrac{\pi}{6} = \boxed{}$$

We use a reference triangle to find $\sin\left(\cos^{-1}\dfrac{5}{13}\right)$.

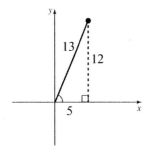

We see that $\sin\left(\cos^{-1}\dfrac{5}{13}\right) = \dfrac{12}{13}$. Now we have

$$\sin\left(\sin^{-1}\dfrac{1}{2}\right)\cos\left(\cos^{-1}\dfrac{5}{13}\right) + \cos\left(\sin^{-1}\dfrac{1}{2}\right)\sin\left(\cos^{-1}\dfrac{5}{13}\right)$$

$$= \dfrac{1}{2}\cdot\boxed{} + \dfrac{\sqrt{3}}{2}\cdot\boxed{}$$

$$= \dfrac{5}{26} + \dfrac{12\sqrt{3}}{26}$$

$$= \dfrac{\boxed{} + 12\sqrt{3}}{\boxed{}}.$$

Section 7.5 Solving Trigonometric Equations

Example 1 Solve: $2\cos x = -1$.

$$2\cos x = -1$$

$$\cos x = \boxed{} \qquad \text{Solving for } \cos x$$

The only two points on the unit circle for which the cosine is $-\dfrac{1}{2}$ are $\dfrac{2\pi}{3}$ and $\boxed{}$. These

numbers plus any multiples of 2π are solutions. Thus, the solutions are

$$\frac{2\pi}{3} + 2k\pi \text{ and } \frac{4\pi}{3} + \boxed{}, \text{ or}$$

$$120° + k \cdot 360° \text{ and } \boxed{}° + k \cdot 360°, \text{ where } k \text{ is any integer.}$$

Example 2 Solve: $4\sin^2 x = 1$.

$$4\sin^2 x = 1$$

$$\sin^2 x = \boxed{}$$

$$\sin x = \pm \frac{1}{2}$$

The points on the unit circle having a sine of $\dfrac{1}{2}$ or $-\dfrac{1}{2}$ are $\dfrac{\pi}{6}$, $\boxed{}$, $\dfrac{7\pi}{6}$, and $\boxed{}$.

These numbers plus any multiple of 2π are solutions. Thus, the solutions are

$$\frac{\pi}{6} + 2k\pi, \quad \frac{5\pi}{6} + 2k\pi, \quad \frac{7\pi}{\boxed{}} + 2k\pi, \text{ and } \frac{11\pi}{6} + 2k\pi,$$

where k is any integer. We can condense the solutions using multiples of $k\pi$. We have

$$\frac{\pi}{6} + k\pi \text{ and } \frac{5\pi}{6} + k\pi,$$

where k is any integer.

Expressed in degrees, the solutions are

$$30° + k \cdot 360°, \quad \boxed{}° + k \cdot 360°, \quad 210° + k \cdot 360°, \text{ and } 330° + k \cdot 360°,$$

where k is any integer. These solutions can be condensed as

$$\boxed{}° + k \cdot 180° \text{ and } 150° + k \cdot 180°,$$

where k is any integer.

Example 3   Solve $3\tan 2x = -3$ in the interval $[0, 2\pi)$.

When solving equations involving double angles, we must be careful to find all solutions in the given interval, in this case $[0, 2\pi)$.

$$3\tan 2x = -3$$

$$\tan 2x = \boxed{}$$

For $0 \le x < 2\pi$, we have $0 \le 2x < 4\pi$ (multiplying by 2), so we solve $\tan 2x = -1$ in $\left[0, \boxed{}\pi\right)$. We find the points $2x$ in $[0, 4\pi)$ for which $\tan 2x = \boxed{}$.

$$2x = \frac{3\pi}{4}, \ \frac{7\pi}{\boxed{}}, \ \frac{\boxed{}\pi}{4}, \text{ and } \frac{15\pi}{4}$$

Dividing by 2, we get the solution in $[0, 2\pi)$:

$$x = \frac{3\pi}{8}, \ \frac{7\pi}{8}, \ \frac{11\pi}{\boxed{}}, \text{ and } \frac{15\pi}{8}.$$

Example 4   Solve $\dfrac{1}{2}\cos\phi + 1 = 1.2108$ in the interval $[0°, 360°)$.

$$\frac{1}{2}\cos\phi + 1 = 1.2108$$

$$\frac{1}{2}\cos\phi = \boxed{}$$

$$\cos\phi = 0.4216$$

Using a calculator set in DEGREE mode, we find the reference angle $\cos^{-1} 0.4216 \approx 65.06°$. Since $\cos\phi$ is positive, the solutions are in quadrants I and $\boxed{}$. The solutions in $[0°, 360°)$ are

$$\boxed{}° \text{ and } 360° - 65.06° = 294.94°.$$

Example 5   Solve $2\cos^2 u = 1 - \cos u$ in the interval $[0°, 360°)$.

$$2\cos^2 u = 1 - \cos u$$

$$2\cos^2 u + \cos u - \boxed{} = 0$$

$$(2\cos u - 1)(\cos u + 1) = \boxed{}$$

$$2\cos u - 1 = 0 \qquad or \quad \cos u + 1 = 0$$

$$2\cos u = \boxed{} \qquad or \qquad \cos u = \boxed{}$$

$$\cos u = \frac{1}{2}$$

$$u = 60°,\ 300° \quad or \qquad\qquad u = 180°$$

The solutions are $60°$, $\boxed{}°$, and $300°$.

Example 6   Solve $\sin^2 \beta - \sin \beta = 0$ in the interval $[0, 2\pi)$.

$$\sin^2 \beta - \sin \beta = 0$$

$$\sin \beta \left(\sin \beta - \boxed{}\right) = 0$$

$$\sin \beta = \boxed{} \quad or \quad \sin \beta - 1 = 0$$

$$\sin \beta = 0 \qquad or \qquad \sin \beta = \boxed{}$$

$$\beta = 0,\ \pi \quad or \qquad \beta = \boxed{}$$

The solutions in $[0, 2\pi)$ are 0, $\dfrac{\pi}{2}$, and $\boxed{}$.

Example 7   Solve $10\sin^2 x - 12\sin x - 7 = 0$ in the interval $[0°, 360°)$.

We will use the quadratic formula with $a = 10$, $b = -12$, and $c = -7$.

$$\sin x = \frac{-b \pm \sqrt{b^2 - 4ac}}{\boxed{}\,a}$$

$$= \frac{-(-12) \pm \sqrt{(-12)^2 - 4 \cdot 10 \cdot (-7)}}{2 \cdot \boxed{}}$$

$$= \frac{12 \pm \sqrt{\boxed{} + 280}}{20} = \frac{12 \pm \sqrt{424}}{\boxed{}}$$

$$\sin x \approx 1.6296 \quad or \quad \sin x \approx -0.4296$$

$-1 \le \sin x \le 1$, so $\sin x \approx 1.6296$ has no solution.

For $\sin x \approx -0.4296$, the reference angle is $25.44°$. Since $\sin x$ is negative, the solutions are in quadrants $\boxed{}$ and IV. The solutions in $[0°, 360°)$ are

$$180° + 25.44° = \boxed{}° \text{ and } \boxed{}° - 25.44° = 334.56°.$$

Example 8   Solve $2\cos^2 x \tan x = \tan x$ in the interval $[0, 2\pi)$.

$$2\cos^2 x \tan x = \tan x$$
$$2\cos^2 x \tan x - \boxed{} = 0$$
$$\tan x \left(2\cos^2 x - \boxed{} \right) = 0$$

$$\tan x = 0 \quad or \quad 2\cos^2 x - 1 = 0$$
$$x = 0, \pi \quad or \quad 2\cos^2 x = \boxed{}$$
$$\cos^2 x = \frac{1}{2}$$
$$\cos x = \pm\sqrt{\frac{1}{2}}$$
$$\cos x = \pm\frac{1}{\sqrt{2}} = \pm\frac{\sqrt{2}}{\boxed{}}$$
$$x = \frac{\pi}{4}, \frac{7\pi}{4}, \frac{\boxed{}\pi}{4}, \frac{5\pi}{\boxed{}}$$

The solutions are $0,$ $\boxed{},$ $\frac{\pi}{4},$ $\frac{7\pi}{4},$ $\frac{3\pi}{\boxed{}},$ and $\frac{\boxed{}\pi}{4}.$

Example 9   Solve $\sin x + \cos x = 1$ in the interval $[0, 2\pi)$.

$$\sin x + \cos x = 1$$
$$(\sin x + \cos x)^2 = 1^2$$
$$\sin^2 x + \boxed{}\sin x \cos x + \cos^2 x = 1$$
$$1 + 2\sin x \cos x = \boxed{} \qquad (\sin^2 x + \cos^2 x = 1)$$
$$2\sin x \cos x = \boxed{}$$
$$\sin 2x = 0 \qquad (2\sin x \cos x = \sin 2x)$$

$$0 \leq x < 2\pi$$

$$0 \leq 2x < 4\pi \qquad \text{Multiplying by 2}$$

$$\sin 2x = 0$$

$$2x = 0,\ \pi,\ 2\pi,\ 3\pi$$

$$x = 0,\ \frac{\pi}{2},\ \boxed{},\ \frac{3\pi}{2}$$

We check the four possible solutions in the original equation, $\sin x + \cos x = 1$.

$$\sin 0 + \cos 0 = \boxed{} + 1 = 1$$

$$\sin \frac{\pi}{2} + \cos \frac{\pi}{2} = \boxed{} + 0 = 1$$

$$\sin \pi + \cos \pi = 0 - \boxed{} = \boxed{}$$

$$\sin \frac{3\pi}{2} + \cos \frac{3\pi}{2} = \boxed{} + 0 = -1$$

We see that 0 and $\frac{\pi}{2}$ check, but $\boxed{}$ and $\frac{3\pi}{2}$ do not. The solutions are 0 and $\frac{\pi}{2}$.

Example 10 Solve $\cos 2x + \sin x = 1$ in the interval $[0, 2\pi)$.

$$\cos 2x + \sin x = 1$$

$$1 - 2\sin^2 x + \sin x = 1 \qquad \left(\cos 2x = 1 - 2\sin^2 x\right)$$

$$-2\sin^2 x + \sin x = \boxed{}$$

$$\sin x \left(-2\sin x + \boxed{}\right) = 0$$

$$\sin x = 0 \qquad or \qquad -2\sin x + 1 = 0$$

$$x = 0,\ \pi \quad or \qquad -2\sin x = \boxed{}$$

$$\sin x = \frac{1}{2}$$

$$x = \frac{\pi}{6},\ \frac{5\pi}{6}$$

All four values check. The solutions are 0, $\frac{\pi}{6}$, $\frac{5\pi}{6}$, and $\boxed{}$.

Example 11 Solve $\tan^2 x + \sec x - 1 = 0$ in the interval $[0, 2\pi)$.

$$\tan^2 x + \sec x - 1 = 0$$

$$\sec^2 x - 1 + \sec x - 1 = 0 \qquad \left(\tan^2 x = \sec^2 x - 1\right)$$

$$\sec^2 x + \sec x - \boxed{} = 0$$

$$\left(\sec x + \boxed{}\right)\left(\sec x - 1\right) = 0$$

$$\sec x + 2 = 0 \qquad or \qquad \sec x - 1 = 0$$

$$\sec x = \boxed{} \qquad or \qquad \sec x = \boxed{}$$

$$\frac{1}{\cos x} = -2 \qquad or \qquad \frac{1}{\cos x} = 1$$

$$-\frac{1}{\boxed{}} = \cos x \qquad or \qquad \boxed{} = \cos x$$

$$x = \frac{2\pi}{3}, \frac{\boxed{}\pi}{3} \qquad or \qquad \boxed{} = x$$

The solutions are 0, $\frac{2\pi}{3}$, and $\boxed{}$.

Example 12 Solve $\sin x - \cos x = \cot x$ in $[0, 2\pi)$.

We use a graphing calculator.

Method 1: Enter $y_1 = \sin x - \cos x$ and $y_2 = 1 / \tan x$ and find the points of intersection of the graphs. A good window to use is $[0, 2\pi, -3, 3]$, $\text{Xscl} = \pi / 2$. Using the INTERSECT feature twice we see that the first coordinates of the points of intersection of the graphs in $[0, 2\pi)$ are approximately 1.1276527 and $\boxed{}$.

Method 2: We start by writing the equation with 0 on one side. We have $\sin x - \cos x - \cot x = 0$. Now graph $y_1 = \sin x - \cos x - 1 / \tan x$ and find the zeros of the function in $[0, 2\pi)$. Using the ZERO feature twice, we find that the zeros are approximately $\boxed{}$ and 5.661357.

Each method leads to the approximate solutions $\boxed{}$ and $\boxed{}$ in $[0, 2\pi)$.

Section 8.1   The Law of Sines

Triangles that are not right triangles are called **oblique** triangles. Any triangle can be solved if at least one side and any other two measures are known. We cannot solve a triangle when only the three angle measures are known, because the angle measures determine only the shape of the triangle and not the size.

The Law of Sines

In any triangle ABC,

$$\frac{a}{\sin A} = \frac{b}{\sin B} = \frac{c}{\sin C}.$$

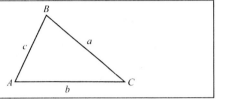

Example 1   In $\triangle EFG$, $e = 4.56$, $E = 43°$, and $G = 57°$. Solve the triangle.

We first make a drawing.

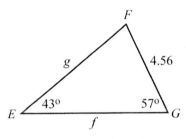

$$F = 180° - \left(43° + \boxed{}°\right) = 80°$$

Now we use the law of sines to find f.

$$\frac{f}{\sin F} = \frac{e}{\sin E}$$

$$\frac{f}{\sin 80°} = \frac{\boxed{}}{\sin 43°}$$

$$f = \frac{4.56 \sin 80°}{\sin 43°} \approx \boxed{}$$

We use the law of sines again to find g.

$$\frac{g}{\sin G} = \frac{\boxed{}}{\sin E}$$

$$\frac{g}{\sin 57°} = \frac{4.56}{\sin 43°}$$

$$g = \frac{\boxed{} \sin 57°}{\sin 43°} \approx 5.61$$

We have solved the triangle.

$$E = 43°, \qquad e \approx \boxed{},$$
$$F = \boxed{}°, \quad f \approx 6.58,$$
$$G = 57°, \qquad g \approx \boxed{}$$

Example 2   *Vietnam Veterans Memorial.* Designed by Maya Lin, the Vietnam Veterans Memorial, in Washington, D.C., consists of two congruent black granite walls on which 58,191 names are inscribed in chronological order of the date of the casualty. Each wall closely approximates a triangle. The height of the memorial at its tallest point is about 120.5 in. The angles formed by the top and the bottom of a wall with the height of the memorial are about 89.4056° and 88.2625°, respectively. Find the lengths of the top and the bottom of a wall rounded to the nearest tenth of an inch.

We make a drawing.

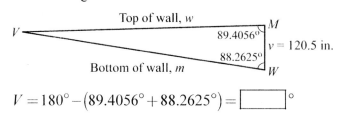

$$V = 180° - (89.4056° + 88.2625°) = \boxed{}°$$

We use the law of sines to find m and w.

$$\frac{m}{\sin M} = \frac{v}{\sin V} \qquad\qquad\qquad \frac{w}{\sin W} = \frac{\boxed{}}{\sin V}$$

$$\frac{\boxed{}}{\sin 89.4056°} = \frac{120.5 \text{ in.}}{\sin 2.3319°} \qquad\qquad \frac{w}{\sin 88.2625°} = \frac{120.5 \text{ in.}}{\sin 2.3319°}$$

$$m = \frac{120.5 \text{ in.}(\sin 89.4056°)}{\sin 2.3319°} \qquad\qquad w = \frac{\boxed{} \text{ in.}(\sin 88.2625°)}{\sin 2.3319°}$$

$$m \approx \boxed{} \text{ in.} \qquad\qquad\qquad w \approx 2960.2 \text{ in.}$$

Thus, $m \approx 2961.4$ in. and $w \approx \boxed{}$ in.

Example 3   In $\triangle QRS$, $q = 15$, $r = 28$, and $Q = 43.6°$. Solve the triangle.

We make a drawing.

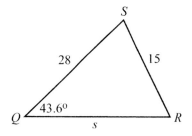

We use the law of sines to find R.

$$\frac{q}{\sin Q} = \frac{r}{\sin R}$$

$$\frac{15}{\sin 43.6°} = \frac{\boxed{}}{\sin R}$$

$$\sin R = \frac{28 \sin 43.6°}{\boxed{}}$$

$$\sin R \approx 1.2873$$

There is no angle with a sine greater than 1, so there is no solution.

Example 4   In $\triangle XYZ$, $x = 23.5$, $y = 9.8$, and $X = 39.7°$. Solve the triangle.

We make a drawing.

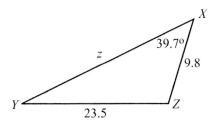

We use the law of sines to find Y.

$$\frac{x}{\sin X} = \frac{y}{\sin Y}$$

$$\frac{\boxed{}}{\sin 39.7°} = \frac{9.8}{\sin Y}$$

$$23.5 \sin Y = 9.8 \sin 39.7°$$

$$\sin Y = \frac{9.8 \sin 39.7°}{\boxed{}}$$

$$\sin Y \approx 0.2664$$

$$Y \approx \boxed{}°, \text{ or}$$

$$Y \approx 180° - 15.4°, \text{ or } 164.6°$$

(There are two angles in $(0°, 180°)$ with a sine of 0.2664.)

We see that an angle of $164.6°$ cannot be an angle of this triangle because the triangle has an angle of $39.7°$ and $39.7° + 164.6° > 180°$. Thus, $Y \approx \boxed{}°$.

$$Z \approx 180° - \left(15.4° + \boxed{}°\right) \approx 124.9°$$

Now we use the law of sines to find z.

$$\frac{z}{\sin Z} = \frac{x}{\sin X}$$

$$\frac{z}{\sin 124.9°} = \frac{\boxed{}}{\sin 39.7°}$$

$$z = \frac{23.5 \sin 124.9°}{\sin 39.7°}$$

$$z \approx \boxed{}$$

We have solved the triangle.

$$X = \boxed{}°, \quad x = \boxed{},$$
$$Y \approx 15.4°, \quad y = 9.8,$$
$$Z \approx 124.9°, \quad z \approx \boxed{}$$

Example 5   In $\triangle ABC$, $b = 15$, $c = 20$, and $B = 29°$. Solve the triangle.

We make a drawing.

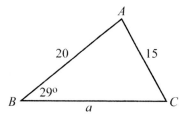

We use the law of sines to find C.

$$\frac{b}{\sin B} = \frac{c}{\sin C}$$

$$\frac{\boxed{}}{\sin 29°} = \frac{\boxed{}}{\sin C}$$

$$\sin C = \frac{20 \sin 29°}{15}$$

$$\sin C \approx 0.6464$$

There are two angles between $0°$ and $180°$ whose sine is 0.6464, so there are _____

no / two

possible solutions of the triangle. We have $C = 40°$ or $C = 140°$.

We find the solution for $C = 40°$.

$$A = 180° - \left(\boxed{}° + 40°\right) = 111°$$

$$\frac{a}{\sin A} = \frac{b}{\sin B}$$

$$\frac{a}{\sin 111°} = \frac{\boxed{}}{\sin 29°}$$

$$a = \frac{15\sin 111°}{\sin 29°} \approx \boxed{}$$

For this situation we have

$$A = 111°, \quad a \approx \boxed{},$$
$$B = 29°, \quad b = 15,$$
$$C = \boxed{}°, \quad c = 20.$$

Now we find the solution for $C = 140°$.

$$A = 180° - \left(29° + \boxed{}°\right) = 11°$$

$$\frac{a}{\sin A} = \frac{b}{\sin B}$$

$$\frac{a}{\sin 11°} = \frac{\boxed{}}{\sin 29°}$$

$$a = \frac{15\sin 11°}{\sin 29°} \approx \boxed{}$$

For this situation we have

$$A = 11°, \quad a \approx \boxed{},$$
$$B = 29°, \quad b = 15,$$
$$C = \boxed{}°, \quad c = 20$$

Example 5 illustrates the **ambiguous case** in which there are two solutions.

The Area of a Triangle

The area K of any $\triangle ABC$ is one-half of the product of the lengths of two sides and the sine of the included angle.

$$K = \frac{1}{2}bc\sin A = \frac{1}{2}ab\sin C = \frac{1}{2}ac\sin B.$$

Example 6   *Area of a Triangular Garden.* A university landscape architecture department is designing a garden for a triangular area in a dormitory complex. Two sides of the garden, formed by the sidewalks in front of buildings A and B, measure 172 ft and 186 ft, respectively, and together form a 53° angle. The third side of the garden, formed by the sidewalk along Crossroads Avenue, measures 160 ft. What is the area of the garden, to the nearest square foot?

We make a drawing.

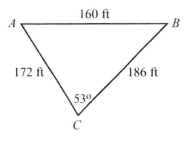

$$K = \frac{1}{2}ab\sin C$$

$$K = \frac{1}{2} \cdot 186 \text{ ft} \cdot \boxed{} \text{ ft} \cdot \sin 53°$$

$$\approx 12,775 \text{ ft}^2$$

The area of the garden is approximately $\boxed{}$ ft².

Section 8.2   The Law of Cosines

The Law of Cosines

In any triangle ABC,

$$a^2 = b^2 + c^2 - 2bc \cos A,$$

$$b^2 = a^2 + c^2 - 2ac \cos B,$$

and $c^2 = a^2 + b^2 - 2ab \cos C.$

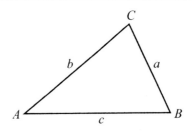

Thus, in any triangle, the square of a side is the sum of the squares of the other two sides, minus twice the product of those sides and the cosine of the included angle. When the included angle is 90°, the law of cosines reduces to the Pythagorean theorem.

Example 1   Solve $\triangle ABC$ if $a = 32$, $b = 71$, and $C = 32.8°$.

We use the law of cosines to find c.

$$c^2 = a^2 + b^2 - 2ab \cos C$$

$$c^2 = 32^2 + 71^2 - 2(32)(\boxed{}) \cos 32.8°$$

$$c^2 \approx 2245.5$$

$$c \approx \boxed{}$$

We could use the law of sines to find a second angle, but the ambiguous case could occur. This is avoided if we use the law of cosines.

$$a^2 = b^2 + c^2 - 2bc \cos A$$

$$32^2 \approx 71^2 + \boxed{}^2 - 2(71)(47) \cos A$$

$$1024 \approx 5041 + 2209 - 6674 \cos A$$

$$1024 \approx 7250 - \boxed{} \cos A$$

$$-6226 \approx -6674 \cos A$$

$$0.9328738 \approx \boxed{}$$

$$21.1° \approx A$$

$$21.1° + B + 32.8° \approx \boxed{}°$$

$$B \approx 126.1°$$

We have solved the triangle.

$A \approx 21.1°,$ $a = \boxed{},$

$B \approx \boxed{}°,$ $b = 71,$

$C = 32.8°,$ $c \approx \boxed{}$

Example 2   *Recording Studio.* A musician is constructing an octagonal recording studio in his home and needs to determine two distances for the electrician. The dimensions for the most acoustically perfect studio are shown in the figure below. Determine the distances from D to F and from D to B to the nearest tenth of an inch.

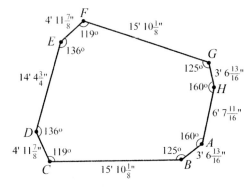

We begin by connecting points D and F to form ΔDEF. We also convert the linear measures to decimal notation in inches.

$$4'11\frac{7}{8}'' = 59.875 \text{ in.}$$

$$14'4\frac{3}{4}'' = 172.75 \text{ in.}$$

We use the law of cosines to fine e.

$e^2 = d^2 + f^2 - 2 \cdot \boxed{} \cdot f \cdot \cos E$

$e^2 = (59.875 \text{ in.})^2 + (172.75 \text{ in.})^2 - 2 \cdot (59.875 \text{ in.})\left(\boxed{}\right)\cos 136°$

$e^2 \approx 48,308.4257 \text{ in}^2$

$e \approx \boxed{}$ in.

Thus, the distance from D to F is approximately $\boxed{}$ in.

Next we connect points D and B to form $\triangle DCB$. We also convert the linear measures to decimal notation in inches.

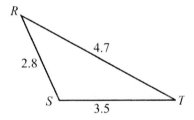

$$4'11\frac{7}{8}'' = 59.875 \text{ in.}$$

$$15'10\frac{1}{8}'' = 190.125 \text{ in.}$$

We use the law of cosines to find c.

$$c^2 = b^2 + d^2 - 2 \cdot b \cdot \boxed{} \cdot \cos C$$

$$c^2 = (59.875 \text{ in.})^2 + (190.125 \text{ in.})^2 - 2(59.875 \text{ in.})\left(\boxed{}\right)\cos 119°$$

$$c^2 \approx 50{,}770.4191 \text{ in}^2$$

$$c \approx \boxed{} \text{ in.}$$

The distance from D to B is approximately $\boxed{}$ in.

Example 3   Solve $\triangle RST$ if $r = 3.5$, $s = 4.7$, and $t = 2.8$.

We make a drawing.

We use the law of cosines to find one of the angle measures. We will find S.

$$s^2 = r^2 + t^2 - 2rt \cos S$$

$$(4.7)^2 = (3.5)^2 + (2.8)^2 - 2(3.5)\left(\boxed{}\right)\cos S$$

$$\cos S = \frac{(3.5)^2 + (2.8)^2 - (4.7)^2}{2(3.5)(2.8)}$$

$$\cos S \approx -0.1020408$$

$$S \approx \boxed{}°$$

Now we find R.

$$r^2 = s^2 + t^2 - 2\boxed{}t\cos R$$

$$(3.5)^2 = (4.7)^2 + (2.8)^2 - 2(4.7)\left(\boxed{}\right)\cos R$$

$$\cos R = \frac{(4.7)^2 + (2.8)^2 - (3.5)^2}{2(4.7)(2.8)}$$

$$\cos R \approx 0.6717325$$

$$R \approx \boxed{}^{\circ}$$

$$T \approx 180° - \left(95.86° + \boxed{}°\right)$$

$$\approx \boxed{}^{\circ}$$

We have solved the triangle.

$$R \approx 47.80°, \qquad r = \boxed{},$$

$$S \approx \boxed{}°, \qquad s = 4.7,$$

$$T \approx 36.34°, \qquad t = \boxed{}$$

Example 4   *Wedge Bevel.* The *bevel* of the wedge (the angle formed at the cutting edge of the wedge) of a log splitter determines the cutting characteristics of the splitter. A small bevel like that of a straight razor makes for a keen edge, but is impractical for heavy-duty cutting because the edge dulls quickly and is prone to chipping. A large bevel is suitable for heavy-duty work like chopping wood. The diagram below illustrates the wedge of a Huskee log splitter. What is its bevel?

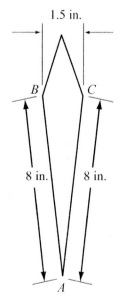

The bevel is angle A. We use the law of cosines to find it.

$$a^2 = b^2 + c^2 - 2bc\cos\boxed{}$$

$$(1.5)^2 = 8^2 + \boxed{}^2 - 2\cdot8\cdot\boxed{}\cos A$$

$$2.25 = 64 + 64 - 128\cos A$$

$$\cos A = \frac{64 + 64 - 2.25}{128}$$

$$\cos A \approx 0.9824$$

$$A \approx \boxed{}^{\circ}$$

The bevel is approximately $\boxed{}°$.

Example 5   In $\triangle ABC$, three measures are given. Determine which law to use when solving the triangle. You need not solve the triangle.

a) $a = 14,\ b = 23,\ c = 10$

b) $a = 207,\ B = 43.8°,\ C = 57.6°$

c) $A = 112°,\ C = 37°,\ a = 84.7$

d) $B = 101°,\ a = 960,\ c = 1042$

e) $b = 17.26,\ a = 27.29,\ A = 39°$

f) $A = 61°,\ B = 39°,\ C = 80°$

a)   We know the lengths of the three sides of the triangle, so we use the law of
$\underline{\hspace{3cm}}$.
      sines / cosines

b)   We know the measures of two angles and the length of the side between them, so we use the law of $\underline{\hspace{3cm}}$.
      sines / cosines

c)   We know the measures of two angles and the length of the side opposite one of them, so we use the law of $\underline{\hspace{3cm}}$.
      sines / cosines

d)   We know the lengths of two sides and the measure of the included angle, so we use the law of $\underline{\hspace{3cm}}$.
      sines / cosines

e)   We know the lengths of two sides and the measure of the angle opposite one of them, so we use the law of $\underline{\hspace{3cm}}$.
      sines / cosines

f)   We know the measures of the three angles. If we know only the angle measures, we cannot determine the lengths of the sides, so this triangle cannot be solved.

Section 8.3   Complex Numbers: Trigonometric Notation

Example 1   Graph each of the following complex numbers.

a) $3+2i$

b) $-4-5i$

c) $-3i$

d) $-1+3i$

e) 2

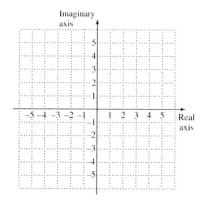

Absolute Value of a Complex Number

The **absolute value of a complex number** $a+bi$ is

$$|a+bi| = \sqrt{a^2 + b^2}.$$

Example 2   Find the absolute value of each of the following.

a) $3+4i$

b) $-2-i$

c) $\dfrac{4}{5}i$

a) $|3+4i| = \sqrt{3^2 + \boxed{}^2} = \sqrt{\boxed{} + 16} = \sqrt{25} = \boxed{}$

b) $|-2-i| = \sqrt{(-2)^2 + (-1)^2} = \sqrt{4 + \boxed{}} = \sqrt{\boxed{}}$

c) $\left|\dfrac{4}{5}i\right| = \left|\boxed{} + \dfrac{4}{5}i\right|$

$$= \sqrt{0^2 + \left(\dfrac{4}{5}\right)^2} = \sqrt{0 + \dfrac{16}{25}}$$

$$= \sqrt{\dfrac{\boxed{}}{25}} = \dfrac{4}{\boxed{}}$$

Trigonometric Notation for Complex Numbers

$a+bi = r(\cos\theta + i\sin\theta)$

Example 3   Find trigonometric notation for each of the following complex numbers.

a) $1+i$ b) $\sqrt{3}-i$

a)   For $1+i$, $a=1$, $b=1$, and $r = \sqrt{1^2+1^2} = \sqrt{\boxed{}}$.

$\sin\theta = \dfrac{b}{r} = \dfrac{1}{\sqrt{2}} = \dfrac{\sqrt{2}}{\boxed{}}$

$\cos\theta = \dfrac{a}{r} = \dfrac{\boxed{}}{\sqrt{2}} = \dfrac{\sqrt{2}}{2}$

The values of the sine and the cosine are both positive, so θ is in quadrant $\boxed{}$. Then

$\theta = \dfrac{\pi}{4}$, or $\boxed{}°$.

$1+i = \sqrt{2}\left(\cos\dfrac{\pi}{4} + \boxed{}\sin\dfrac{\pi}{4}\right)$, or

$1+i = \sqrt{2}\left(\cos 45° + i\sin\boxed{}°\right)$

b)   For $\sqrt{3}-i$, $a=\sqrt{3}$, $b=\boxed{}$, and $r = \sqrt{\left(\sqrt{3}\right)^2+(-1)^2} = \sqrt{3+\boxed{}} = \sqrt{4} = \boxed{}$.

$\sin\theta = \dfrac{b}{r} = \dfrac{-1}{2} = -\boxed{}$

$\cos\theta = \dfrac{a}{r} = \dfrac{\sqrt{3}}{\boxed{}}$

We see that $\sin\theta$ is $\underline{}$ and $\cos\theta$ is $\underline{}$, so θ is in
$\qquad$ positive / negative $\qquad\qquad$ positive / negative

quadrant $\boxed{}$. Then $\theta = \dfrac{11\pi}{6}$, or $\boxed{}°$.

$\sqrt{3}-i = 2\left(\cos\dfrac{11\pi}{6} + \boxed{}\sin\dfrac{11\pi}{6}\right)$, or

$\sqrt{3}-i = \boxed{}\left(\cos 330° + i\sin 330°\right)$

Example 4   Find standard notation $a + bi$, for each of the following complex numbers.

a) $2(\cos 120° + i \sin 120°)$

b) $\sqrt{8}\left(\cos \dfrac{7\pi}{4} + i \sin \dfrac{7\pi}{4}\right)$

a) $2(\cos 120° + i \sin 120°) = 2 \cos 120° + (2 \sin 120°)i$

$$a = r\cos\theta = 2\cos 120° = 2\left(\boxed{}\right) = \boxed{}$$

$$b = r\sin\theta = 2\sin 120° = \boxed{}\left(\dfrac{\sqrt{3}}{2}\right) = \boxed{}$$

$$2(\cos 120° + i \sin 120°) = \boxed{} + \sqrt{3}\,\boxed{}$$

b) $\sqrt{8}\left(\cos \dfrac{7\pi}{4} + i \sin \dfrac{7\pi}{4}\right) = \sqrt{8}\cos\dfrac{7\pi}{4} + \left(\sqrt{8}\sin\dfrac{7\pi}{4}\right)\boxed{}$

$$a = r\cos\theta = \sqrt{8}\cos\dfrac{7\pi}{4} = \sqrt{8}\left(\dfrac{\sqrt{2}}{2}\right) = \dfrac{\sqrt{16}}{\boxed{}} = 2$$

$$b = r\sin\theta = \sqrt{8}\sin\dfrac{7\pi}{4} = \sqrt{8}\left(-\dfrac{\sqrt{2}}{2}\right) = -\dfrac{\sqrt{\boxed{}}}{2} = -2$$

$$\sqrt{8}\left(\cos\dfrac{7\pi}{4} + i\sin\dfrac{7\pi}{4}\right) = \boxed{} - \boxed{}\,i$$

Complex Numbers: Multiplication

For any complex numbers $r_1(\cos\theta_1 + i\sin\theta_1)$ and $r_2(\cos\theta_2 + i\sin\theta_2)$,

$$r_1(\cos\theta_1 + i\sin\theta_1)\cdot r_2(\cos\theta_2 + i\sin\theta_2)$$
$$= r_1 r_2\left[\cos(\theta_1 + \theta_2) + i\sin(\theta_1 + \theta_2)\right].$$

Example 5   Multiply and express the answer to each of the following in standard notation.

a) $3(\cos 40° + i\sin 40°)$ and $4(\cos 20° + i\sin 20°)$

b) $2(\cos\pi + i\sin\pi)$ and $3\left[\cos\left(-\dfrac{\pi}{2}\right) + i\sin\left(-\dfrac{\pi}{2}\right)\right]$

a) $3(\cos 40° + i\sin 40°)\cdot 4(\cos 20° + i\sin 20°)$

$$= 3\cdot\boxed{}\left(\cos(40° + 20°) + i\sin\left(40° + \boxed{}°\right)\right)$$

$$= \boxed{}\left(\cos 60° + i\sin\boxed{}°\right)$$

$$= 12\left(\dfrac{1}{2} + i\cdot\dfrac{\sqrt{3}}{2}\right) = \boxed{} + \boxed{}\sqrt{3}\,i$$

b) $2(\cos\pi+i\sin\pi)\cdot 3\left[\cos\left(-\dfrac{\pi}{2}\right)+i\sin\left(-\dfrac{\pi}{2}\right)\right]$

$=\boxed{}\cdot 3\left[\cos\left(\pi+\left(-\dfrac{\pi}{2}\right)\right)+i\sin\left(\boxed{}+\left(-\dfrac{\pi}{2}\right)\right)\right]$

$=\boxed{}\left(\cos\dfrac{\pi}{2}+i\sin\dfrac{\pi}{2}\right)$

$=6\left(\boxed{}+i\cdot\boxed{}\right)=0+6i=\boxed{}i$

Example 6   Convert to trigonometric notation and multiply: $(1+i)\left(\sqrt{3}-i\right).$

$1+i=\sqrt{2}\left(\cos 45°+i\sin 45°\right)$ (See Example 3(a).)

$\sqrt{3}-i=2\left(\cos 330°+i\sin 330°\right)$ (See Example 3(b).)

$\sqrt{2}\left(\cos 45°+i\sin 45°\right)\cdot\boxed{}\left(\cos 330°+i\sin 330°\right)$

$=\boxed{}\sqrt{2}\left(\cos\left(45°+\boxed{}°\right)+i\sin\left(\boxed{}°+330°\right)\right)$

$=2\sqrt{2}\left(\cos 375°+i\sin 375°\right)$

$=2\sqrt{2}\left(\cos 15°+i\sin\boxed{}°\right)$ (375° and 15° are coterminal.)

Complex Numbers: Division

For any complex numbers $r_1\left(\cos\theta_1+i\sin\theta_1\right)$ and $r_2\left(\cos\theta_2+i\sin\theta_2\right),\ r_2\neq 0,$

$$\dfrac{r_1\left(\cos\theta_1+i\sin\theta_1\right)}{r_2\left(\cos\theta_2+i\sin\theta_2\right)}=\dfrac{r_1}{r_2}\left[\cos\left(\theta_1-\theta_2\right)+i\sin\left(\theta_1-\theta_2\right)\right].$$

Example 7   Divide $2\left(\cos\dfrac{3\pi}{2}+i\sin\dfrac{3\pi}{2}\right)$ by $4\left(\cos\dfrac{\pi}{2}+i\sin\dfrac{\pi}{2}\right)$ and express the solution in standard notation.

$$\frac{2\left(\cos\dfrac{3\pi}{2}+i\sin\dfrac{3\pi}{2}\right)}{4\left(\cos\dfrac{\pi}{2}+i\sin\dfrac{\pi}{2}\right)}=\frac{2}{\boxed{}}\left[\cos\left(\frac{3\pi}{2}-\frac{\pi}{2}\right)+i\sin\left(\frac{3\pi}{2}-\boxed{}\right)\right]$$

$$=\frac{1}{2}\left(\cos\pi+i\sin\boxed{}\right)$$

$$=\frac{1}{2}\left(-1+i\cdot\boxed{}\right)$$

$$=-\frac{1}{2}$$

Example 8 Convert to trigonometric notation and divide: $\dfrac{1+i}{1-i}$.

$1+i = \sqrt{2}\left(\cos 45° + i\sin 45°\right)$ (See Example 3(a).)

$1-i = \sqrt{2}\left(\cos 315° + i\sin 315°\right)$

$\dfrac{1+i}{1-i} = \dfrac{\sqrt{2}\left(\cos 45° + i\sin 45°\right)}{\sqrt{2}\left(\cos 315° + i\sin 315°\right)}$

$\qquad = \dfrac{\sqrt{2}}{\sqrt{2}}\left(\cos\left(45° - \boxed{}°\right) + i\sin\left(\boxed{}° - 315°\right)\right)$

$\qquad = \cos\left(-270°\right) + \boxed{}\sin\left(-270°\right)$

$\qquad = \boxed{} + i\boxed{}$

$\qquad = \boxed{}$

DeMoivre's Theorem

For any complex number $r\left(\cos\theta + i\sin\theta\right)$ and any natural number n,

$\left[r\left(\cos\theta + i\sin\theta\right)\right]^{n} = r^{n}\left(\cos n\theta + i\sin n\theta\right).$

Example 9 Find each of the following.

a) $(1+i)^{9}$

b) $\left(\sqrt{3}-i\right)^{10}$

a) $(1+i)^{9} = \left[\sqrt{2}\left(\cos 45° + i\sin 45°\right)\right]^{9}$

$\qquad = \left(\sqrt{2}\right)^{9}\left[\cos\left(\boxed{}\cdot 45°\right) + i\sin\left(\boxed{}\cdot 45°\right)\right]$

$\qquad = 2^{9/2}\left(\cos 405° + i\sin\boxed{}°\right)$

$\qquad = \boxed{}\sqrt{2}\left(\dfrac{\sqrt{2}}{2} + i\dfrac{\sqrt{2}}{\boxed{}}\right)$

$\qquad = \boxed{} + \boxed{}\,i$

b) $\left(\sqrt{3}-i\right)^{10} = \left[2\left(\cos 330° + i\sin 330°\right)\right]^{10}$

$\qquad = 2^{10}\left(\cos 3300° + \boxed{}\sin 3300°\right)$

$\qquad = 1024\left(\cos 60° + i\sin \boxed{}°\right)$

$\qquad = 1024\left(\boxed{} + i\dfrac{\sqrt{3}}{2}\right)$

$\qquad = \boxed{} + 512\sqrt{3}\,\boxed{}$

Roots of Complex Numbers

The nth roots of a complex number $r\left(\cos\theta + i\sin\theta\right)$, $r \neq 0$, are given by

$$r^{1/n}\left[\cos\left(\dfrac{\theta}{n} + k\cdot\dfrac{360°}{n}\right) + i\sin\left(\dfrac{\theta}{n} + k\cdot\dfrac{360°}{n}\right)\right],$$

where $k = 0, 1, 2, \ldots, n-1$.

Example 10   Find the square roots of $2 + 2\sqrt{3}\,i$.

$$2 + 2\sqrt{3}\,i = 4\left(\cos 60° + i\sin 60°\right)$$

$$\left[4\left(\cos 60° + i\sin 60°\right)\right]^{1/2}$$

$$= \boxed{}^{1/2}\left[\cos\left(\dfrac{60°}{\boxed{}} + k\cdot\dfrac{360°}{2}\right) + \sin\left(\dfrac{60°}{2} + \boxed{}\cdot\dfrac{360°}{2}\right)\right]$$

$$= 2\left[\cos\left(30° + k\cdot\boxed{}°\right) + i\sin\left(\boxed{}° + k\cdot 180°\right)\right], \quad k = 0, 1$$

For $k = 0$:

$$2\left[\cos\left(30° + \boxed{}\cdot 180°\right) + i\sin\left(30° + \boxed{}\cdot 180°\right)\right]$$

$$= 2\left(\cos 30° + i\sin 30°\right) = \boxed{}\left[\dfrac{\sqrt{3}}{2} + i\left(\dfrac{1}{2}\right)\right]$$

$$= \sqrt{3} + \boxed{}$$

For $k = 1$:

$$2\left[\cos\left(30° + \boxed{}\cdot 180°\right) + i\sin\left(30° + \boxed{}\cdot 180°\right)\right]$$

$$= \boxed{}\left(\cos 210° + i\sin \boxed{}°\right)$$

$$= -\sqrt{3} - \boxed{}$$

Example 11 Find the cube roots of 1. Then locate them on a graph.

$$1 = 1(\cos 0° + i \sin 0°)$$

$$\left[1(\cos 0° + i \sin 0°)\right]^{1/3}$$

$$= \boxed{}^{1/3}\left[\cos\left(\frac{0°}{\boxed{}} + k \cdot \frac{360°}{\boxed{}}\right) + i \sin\left(\frac{0°}{3} + \boxed{} \cdot \frac{360°}{3}\right)\right]$$

$$= \cos(k \cdot 120°) + i \sin\left(\boxed{} \cdot 120°\right), \ k = 0, 1, 2$$

For $k = 0$:

$$\cos(0 \cdot 120°) + i \sin\left(\boxed{} \cdot 120°\right)$$

$$= \cos 0° + \boxed{} \sin 0° = \boxed{}$$

For $k = 1$:

$$\cos(1 \cdot 120°) + i \sin\left(\boxed{} \cdot 120°\right)$$

$$= \cos 120° + i \sin \boxed{}° = -\frac{1}{2} + \frac{\sqrt{3}}{2}i$$

For $k = 2$:

$$\cos\left(\boxed{} \cdot 120°\right) + \boxed{} \sin(2 \cdot 120°)$$

$$= \cos 240° + i \sin 240° = -\frac{1}{2} - \frac{\sqrt{3}}{2}\boxed{}$$

The cube roots are evenly spaced about a circle with radius 1.

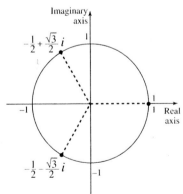

Example 12   Solve: $x^5 + i = 0$.

$$x^5 + i = 0$$

$$x^5 = -i$$

$$x^5 = 1(\cos 270° + i \sin 270°)$$

$$x = \sqrt[5]{1}\left(\cos\left(\frac{270°}{\boxed{}} + k \cdot \frac{360°}{5}\right) + i \sin\left(\frac{270°}{5} + \boxed{} \cdot \frac{360°}{5}\right)\right), \; k = 0, 1, 2, 3, 4$$

$$\frac{270°}{5} = 54°, \quad \frac{360°}{5} = \boxed{}°$$

$$x_1 = 1(\cos 54° + i \sin 54°)$$

$$x_2 = 1\left(\cos 126° + \boxed{}\, \sin 126°\right)$$

$$x_3 = 1\left(\cos 198° + i \sin \boxed{}\,°\right)$$

$$x_4 = 1\left(\cos \boxed{}\,° + i \sin 270°\right) = -i$$

$$x_5 = 1(\cos 342° + \sin 342°)$$

The solutions are

$$\cos 54° + i \sin 54°,$$

$$\cos 126° + i \sin \boxed{}\,°,$$

$$\cos \boxed{}\,° + i \sin 198°,$$

$$\cos 270° + \boxed{}\, \sin 270°, \text{ or } -i, \text{ and}$$

$$\cos 342° + i \sin 342°.$$

Section 8.4   Polar Coordinates and Graphs

On a polar graph, the origin is called the **pole**, and the positive half of the *x*-axis is called the **polar axis**. A point in the polar coordinate system has coordinates (r, θ).

To plot points on a polar graph:

1. Locate the directed angle θ from the polar axis to the ray OP.
2. Move a directed distance r from the pole. If $r > 0$, move along ray OP. If $r < 0$, move in the opposite direction of ray OP.

Example 1   Graph each of the following points.

a) $A(3, 60°)$

b) $B(0, 10°)$

c) $C(-5, 120°)$

d) $D(1, -60°)$

e) $E\left(2, \dfrac{3\pi}{2}\right)$

f) $F\left(-4, \dfrac{\pi}{3}\right)$

g) $G\left(-5, -\dfrac{\pi}{4}\right)$

h) $H\left(5, \dfrac{3\pi}{4}\right)$

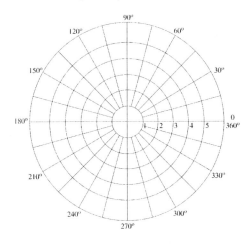

To convert from rectangular coordinates to polar coordinates and from polar coordinates to rectangular coordinates, we need to recall the following relationships.

$$r = \sqrt{x^2 + y^2}$$

$$\cos\theta = \frac{x}{r}, \text{ or } x = r\cos\theta$$

$$\sin\theta = \frac{y}{r}, \text{ or } y = r\sin\theta$$

$$\tan\theta = \frac{y}{x}$$

Example 2   Convert each of the following to polar coordinates.

a) $(3, 3)$ b) $\left(2\sqrt{3}, -2\right)$

a) $r = \sqrt{3^2 + 3^2} = \sqrt{\boxed{}} = \boxed{}\sqrt{2}$

$\tan \theta = \dfrac{3}{3} = \boxed{}$ and $(3, 3)$ is in quadrant $\boxed{}$, so $\theta = 45°$, or $\dfrac{\pi}{4}$.

Thus, $(r, \theta) = \left(3\sqrt{2}, \boxed{}°\right)$, or $\left(3\sqrt{2}, \dfrac{\pi}{4}\right)$.

There are many other possibilities for polar coordinates including

$\left(3\sqrt{2}, -315°\right)$, or $\left(-3\sqrt{2}, \dfrac{5\pi}{\boxed{}}\right)$.

b) $r = \sqrt{\left(2\sqrt{3}\right)^2 + (-2)^2} = \sqrt{12 + \boxed{}} = \sqrt{16} = \boxed{}$

$\tan \theta = \dfrac{-2}{2\sqrt{3}} = -\dfrac{1}{\sqrt{3}}$ and $\left(2\sqrt{3}, -2\right)$ is in quadrant $\boxed{}$, so $\theta = 330°$, or $\dfrac{11\pi}{\boxed{}}$.

Thus, $(r, \theta) = (4, 330°)$, or $\left(\boxed{}, \dfrac{11\pi}{6}\right)$. Other possibilities include

$\left(4, -\dfrac{\pi}{6}\right)$ and $\left(\boxed{}, 150°\right)$.

Example 3   Convert each of the following to rectangular coordinates.

a) $\left(10, \dfrac{\pi}{3}\right)$ b) $(-5, 135°)$

a) $r = 10, \; \theta = \dfrac{\pi}{3}$

$x = r\cos\theta = \boxed{}\cos\dfrac{\pi}{3} = 10 \cdot \dfrac{1}{2} = \boxed{}$

$y = r\sin\theta = 10\sin\dfrac{\pi}{3} = 10 \cdot \dfrac{\sqrt{3}}{\boxed{}} = \boxed{}\sqrt{3}$

Thus, $\left(10, \dfrac{\pi}{3}\right) = \left(\boxed{}, 5\sqrt{3}\right)$.

b) $r = -5, \ \theta = 135°$

$$x = r\cos\theta = \boxed{}\cos 135° = -5\left(-\frac{\sqrt{2}}{2}\right) = \frac{5\sqrt{2}}{\boxed{}}$$

$$y = r\sin\theta = -5\sin\boxed{}° = -5\left(\frac{\sqrt{2}}{2}\right) = -\frac{\boxed{}\sqrt{2}}{2}$$

Thus, $(-5, 135°) = \left(\dfrac{5\sqrt{2}}{2}, \ -\dfrac{5\sqrt{2}}{\boxed{}}\right).$

Example 4   Convert each of the following to a polar equation.

a) $x^2 + y^2 = 25$

b) $2x - y = 5$

a)   Recall that $x = r\cos\theta$ and $y = \boxed{}\sin\theta$.

$$x^2 + y^2 = 25$$
$$(r\cos\theta)^2 + (r\sin\theta)^2 = \boxed{}$$
$$r^2\cos^2\theta + r^2\sin^2\theta = 25$$
$$\boxed{}(\cos^2\theta + \sin^2\theta) = 25$$
$$r^2 = \boxed{} \qquad\qquad (\cos^2\theta + \sin^2\theta = 1)$$
$$r = \boxed{}$$

b)
$$2x - y = 5$$
$$\boxed{}(r\cos\theta) - r\sin\theta = \boxed{}$$
$$r(2\cos\theta - \sin\theta) = 5$$

Example 5   Convert each of the following to a rectangular equation.

a) $r = 4$ b) $r\cos\theta = 6$ c) $r = 2\cos\theta + 3\sin\theta$

a) $r = 4$

$$\sqrt{x^2 + y^2} = \boxed{}$$ $r = \sqrt{x^2 + y^2}$

$$x^2 + y^2 = \boxed{}$$ Squaring both sides

b) $r\cos\theta = 6$

$$x = \boxed{}$$ $\left(x = r\cos\theta\right)$

c) $r = 2\cos\theta + 3\sin\theta$

$$r^2 = 2r\cos\theta + 3\boxed{}\sin\theta$$ Multiplying by r

$$x^2 + y^2 = 2\boxed{} + \boxed{}\,y$$

Example 6   Graph: $r = 1 - \sin\theta$.

Make a table of values, plot these points, and draw the curve.

θ	r
0°	1
30°	$\boxed{}$
90°	0
150°	$\boxed{}$
180°	$\boxed{}$
210°	1.5
270°	$\boxed{}$
330°	$\boxed{}$
360°	1
390°	0.5

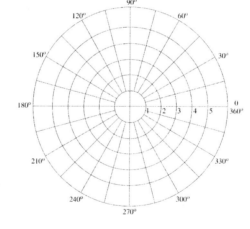

The heart-shaped graph is called a *cardioid*.

Example 7   Graph each of the following polar equations. Try to visualize the shape of the curve before graphing it.

a) $r = 3$ b) $r = 5\sin\theta$ c) $r = 2\csc\theta$

a)   For any θ, $r = \boxed{}$ so the graph will be all points 3 units from the origin. This is a circle with radius $\boxed{}$ centered at the origin.

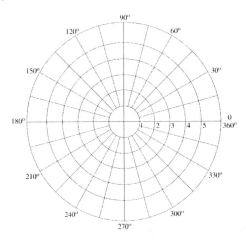

b) $r = 5\sin\theta$

$\qquad\qquad r^2 = 5\boxed{}\sin\theta$ Multiplying by r

$\qquad x^2 + y^2 = 5y$

$x^2 + y^2 - 5y = \boxed{}$

The graph of this equation is a circle. Find some points on the graph, plot them, and draw the graph.

θ	r
0°	0
30°	$\boxed{}$
90°	5
150°	$\boxed{}$
180°	$\boxed{}$

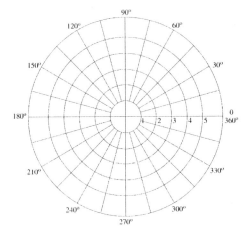

c) $r = 2\csc\theta$

$r = \dfrac{\boxed{}}{\sin\theta}$

$r\sin\theta = 2$

$y = \boxed{}$

The graph is a horizontal line $\boxed{}$ units _____ the origin.
$\underset{\text{above / below}}{}$

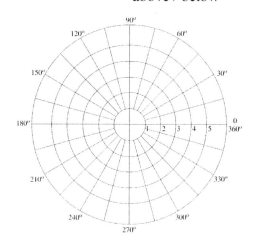

Example 8 Graph the equation $r + 1 = 2\cos 2\theta$ with a graphing calculator.

We start by solving for r. We have

$r = 2\cos 2\theta - \boxed{}$.

With the calculator in POLAR mode, enter and graph $\boxed{} = 2\cos 2\theta - 1$. A good choice for the viewing window is $[-6, 6, -4, 4]$. (Sketch the graph in the space below.)

Section 8.5 Vectors and Applications

> **Vector**
>
> A **vector** in the plane is a directed line segment. Two vectors are **equivalent** if they have the same *magnitude* and the same *direction*.

Example 1   The vectors **u**, $\overline{OR}$, and **w** are shown in the following figure. Show that $\mathbf{u} = \overline{OR} = \mathbf{w}$.

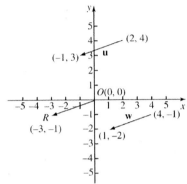

We first find the length, or magnitude, of each vector.

$$|\mathbf{u}| = \sqrt{\left(2-(-1)\right)^2 + \left(4-\boxed{}\right)^2} = \sqrt{3^2 + 1^2} = \sqrt{9+1} = \sqrt{\boxed{}}$$

$$|\overline{OR}| = \sqrt{\left(0-(-3)\right)^2 + \left(\boxed{}-(-1)\right)^2} = \sqrt{3^2+1^2} = \sqrt{\boxed{}+1} = \sqrt{10}$$

$$|\mathbf{w}| = \sqrt{\left(4-\boxed{}\right)^2 + \left(-1-(-2)\right)^2} = \sqrt{3^2+1^2} = \sqrt{9+\boxed{}} = \sqrt{\boxed{}}$$

The vectors have the same magnitude. Now we find the slopes.

$$\text{Slope of } \mathbf{u} = \frac{4-\boxed{}}{2-(-1)} = \frac{1}{\boxed{}}$$

$$\text{Slope of } \overline{OR} = \frac{\boxed{}-(-1)}{0-(-3)} = \frac{1}{\boxed{}}$$

$$\text{Slope of } |\mathbf{w}| = \frac{\boxed{}-(-2)}{4-\boxed{}} = \frac{\boxed{}}{3}$$

The vectors have the same direction.

Since **u**, $\overline{OR}$, and **w** have the same magnitude and the same direction, $\mathbf{u} = \overline{OR} = \mathbf{w}$.

> If two forces act on an object, the combined effect is the **sum**, or **resultant**, of the two forces.

Example 2   Forces of 15 newtons and 25 newtons act on an object at right angles to each other. Find their sum, or resultant, giving the magnitude of the resultant and the angle that it makes with the larger force.

We make a drawing, letting **v** represent the resultant.

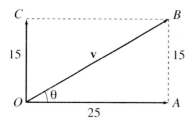

$$|\mathbf{v}|^2 = 25^2 + 15^2$$

$$|\mathbf{v}| = \sqrt{25^2 + 15^2} \approx \boxed{} \text{ newtons}$$

$$\tan\theta = \frac{15}{\boxed{}} = 0.6$$

$$\theta \approx \boxed{}^{\circ}$$

The resultant has a magnitude of $\boxed{}$ newtons and makes an angle of about $\boxed{}^{\circ}$ with the larger force.

Example 3   An airplane travels on a bearing of $100°$ at an airspeed of 190 km/h while a wind is blowing 48 km/h from $220°$. Find the ground speed of the airplane and the direction of its track, or course, over the ground.

We make a drawing.

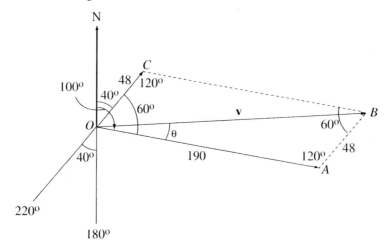

Consider $\triangle OBA$. We use the law of cosines to find $\left|\mathbf{v}\right|$.

$$\left|\mathbf{v}\right|^2 = 48^2 + 190^2 - 2 \cdot \boxed{} \cdot 190 \cos 120° = 47{,}524$$
$$\left|\mathbf{v}\right| = 218 \text{ km/h}$$

Now we use the law of sines to find θ.

$$\frac{48}{\sin\theta} = \frac{\boxed{}}{\sin 120°}$$

$$\sin\theta = \frac{\boxed{}\sin 120°}{218} \approx \boxed{}$$

$$\theta \approx 11°$$

The ground speed of the airplane is $\boxed{}$ km/h, and its track is in the direction of $100° - 11°$, or $\boxed{}°$.

Components

Given a vector $\mathbf{w}$, we may want to find two other vectors $\mathbf{u}$ and $\mathbf{v}$ whose sum is $\mathbf{w}$. The vectors $\mathbf{u}$ and $\mathbf{v}$ are called **components** of $\mathbf{w}$ and the process of finding them is called **resolving**, or **representing**, a vector into its vector components. We usually look for perpendicular components, most often parallel to the *x*- and *y*-axes. These are called the **horizontal component** and the **vertical component** of a vector.

Example 4   A vector $\mathbf{w}$ has a magnitude of 130 and is inclined $40°$ with the horizontal. Resolve the vector into its horizontal component and its vertical component.

We make a drawing.

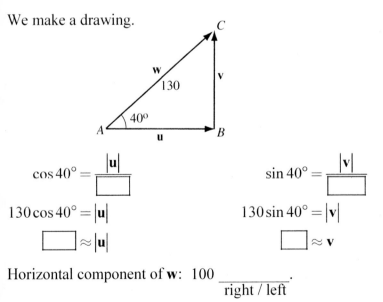

$$\cos 40° = \frac{\left|\mathbf{u}\right|}{\boxed{}} \qquad\qquad \sin 40° = \frac{\left|\mathbf{v}\right|}{\boxed{}}$$

$$130\cos 40° = \left|\mathbf{u}\right| \qquad\qquad 130\sin 40° = \left|\mathbf{v}\right|$$

$$\boxed{} \approx \left|\mathbf{u}\right| \qquad\qquad \boxed{} \approx \mathbf{v}$$

Horizontal component of $\mathbf{w}$: 100 $\underset{\text{right / left}}{\underline{}}$.

Vertical component of $\mathbf{w}$: 84 $\underset{\text{up / down}}{\underline{}}$

Example 5 *Shipping Crate.* A wooden shipping crate that weighs 816 lb is placed on a loading ramp that makes an angle of $25°$ with the horizontal. To keep the crate from sliding, a chain is hooked to the crate and to a pole at the top of the ramp. Find the magnitude of the components of the crate's weight (disregarding friction) perpendicular to and parallel to the incline.

We make a drawing.

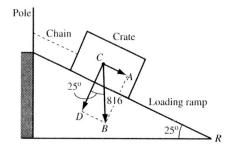

$\left|\overrightarrow{CB}\right|$ = the weight of the crate = 816 lb,

$\left|\overrightarrow{CD}\right|$ = the magnitude of the component _____ to the ramp, and
  perpendicular / parallel

$\left|\overrightarrow{CA}\right|$ = the magnitude of the component _____ to the ramp.
  perpendicular / parallel

We have $\angle DCB = \angle R = 25°$ because corresponding sides of the two angles are perpendicular.

$$\cos 25° = \frac{\left|\overrightarrow{CD}\right|}{\boxed{}}$$

$$\left|\overrightarrow{CD}\right| = 816 \cos 25°$$

$$\approx \boxed{} \text{ lb}$$

$$\sin 25° = \frac{\left|\overrightarrow{CA}\right|}{816}$$

$$\left|\overrightarrow{CA}\right| = \boxed{} \sin 25°$$

$$\approx \boxed{} \text{ lb}$$

Section 8.6   Vector Operations

Component Form of a Vector

The **component form** of $\overline{AC}$ with $A = (x_1, y_1)$ and $C = (x_2, y_2)$ is

$$\overline{AC} = \langle x_2 - x_1, y_2 - y_1 \rangle.$$

Example 1   Find the component form of $\overline{CF}$ if $C = (-4, -3)$ and $F = (1, 5)$.

$$\overline{CF} = \left\langle \boxed{} - (-4), \boxed{} - (-3) \right\rangle = \left\langle 5, \boxed{} \right\rangle$$

Note that $\overline{CF}$ is equivalent to the position vector $\overline{OP}$ with $P = (5, 8)$.

Length of a Vector

The **length**, or **magnitude**, of a vector $\mathbf{v} = \langle v_1, v_2 \rangle$ is given by

$$|\mathbf{v}| = \sqrt{v_1^2 + v_2^2}.$$

Example 2   Find the length, or magnitude, of vector $\mathbf{v} = \langle 5, 8 \rangle$.

$$|\mathbf{v}| = \sqrt{v_1^2 + v_2^2} = \sqrt{5^2 + \boxed{}^2}$$

$$= \sqrt{\boxed{} + 64} = \sqrt{\boxed{}}$$

Scalar Multiplication

For a real number k and a vector $\mathbf{v} = \langle v_1, v_2 \rangle$, the **scalar product** of k and $\mathbf{v}$ is

$$k\mathbf{v} = k\langle v_1, v_2 \rangle = \langle kv_1, kv_2 \rangle.$$

The vector $k\mathbf{v}$ is a **scalar multiple** of the vector $\mathbf{v}$.

Example 3   Let $\mathbf{u} = \langle -5, 4 \rangle$ and $\mathbf{w} = \langle 1, -1 \rangle$. Find $-7\mathbf{w}$, $3\mathbf{u}$, and $-1\mathbf{w}$.

$$-7\mathbf{w} = \boxed{}\langle 1, -1 \rangle = \left\langle -7, \boxed{} \right\rangle$$

$$3\mathbf{u} = \boxed{}\langle -5, 4 \rangle = \left\langle \boxed{}, 12 \right\rangle$$

$$-1\mathbf{w} = \boxed{}\langle 1, -1 \rangle = \left\langle -1, \boxed{} \right\rangle$$

Vector Addition and Vector Subtraction

If $\mathbf{u} = \langle u_1, u_2 \rangle$ and $\mathbf{v} = \langle v_1, v_2 \rangle$, then

$$\mathbf{u} + \mathbf{v} = \langle u_1 + v_1, u_2 + v_2 \rangle \text{ and } \mathbf{u} - \mathbf{v} = \langle u_1 - v_1, u_2 - v_2 \rangle.$$

Example 4 Perform the following calculations, where $\mathbf{u} = \langle 7, 2 \rangle$ and $\mathbf{v} = \langle -3, 5 \rangle$.

a) $\mathbf{u} + \mathbf{v}$ b) $\mathbf{u} - 6\mathbf{v}$

c) $3\mathbf{u} + 4\mathbf{v}$ d) $|5\mathbf{v} - 2\mathbf{u}|$

a) $\mathbf{u} + \mathbf{v} = \langle 7, \boxed{} \rangle + \langle \boxed{}, 5 \rangle = \langle \boxed{} + (-3), 2 + \boxed{} \rangle = \langle 4, \boxed{} \rangle$

b) $\mathbf{u} - 6\mathbf{v} = \langle 7, 2 \rangle - \boxed{} \langle -3, 5 \rangle = \langle 7, 2 \rangle - \langle -18, \boxed{} \rangle$

$= \langle \boxed{} - (-18), 2 - \boxed{} \rangle = \langle 25, \boxed{} \rangle$

c) $3\mathbf{u} + 4\mathbf{v} = \boxed{} \langle 7, 2 \rangle + 4 \langle -3, \boxed{} \rangle = \langle 21, \boxed{} \rangle + \langle \boxed{}, 20 \rangle$

$= \langle 21 + (-12), 6 + \boxed{} \rangle = \langle \boxed{}, 26 \rangle$

d) $|5\mathbf{v} - 2\mathbf{u}| = |5 \langle -3, \boxed{} \rangle - 2 \langle \boxed{}, 2 \rangle|$

$= |\langle -15, \boxed{} \rangle - \langle \boxed{}, 4 \rangle|$

$= |\langle -29, 21 \rangle| = \sqrt{(-29)^2 + \boxed{}^2}$

$= \sqrt{1282} \approx \boxed{}$

Unit Vector

A vector with length, or magnitude, 1 is a **unit vector**.

If $\mathbf{v}$ is a vector and $\mathbf{v} \neq$ the zero vector $\langle 0, 0 \rangle$, then

$$\frac{1}{|\mathbf{v}|} \cdot \mathbf{v}, \text{ or } \frac{\mathbf{v}}{|\mathbf{v}|},$$

is a unit vector in the direction of $\mathbf{v}$.

Example 5   Find a unit vector that has the same direction as the vector $\mathbf{w} = \langle -3, 5 \rangle$.

$$|\mathbf{w}| = \sqrt{(-3)^2 + \boxed{}^2} = \sqrt{\boxed{} + 25} = \sqrt{\boxed{}}$$

$$\mathbf{u} = \frac{1}{\sqrt{34}}\mathbf{w} = \frac{1}{\sqrt{34}}\langle -3, 5 \rangle = \left\langle -\frac{\boxed{}}{\sqrt{34}}, \frac{\boxed{}}{\sqrt{34}} \right\rangle$$

Unit vectors parallel to the *x*- and *y*-axes are defined as $\mathbf{i} = \langle 1, 0 \rangle$ and $\mathbf{j} = \langle 0, 1 \rangle$, respectively. Any vector $\mathbf{v} = \langle v_1, v_2 \rangle$ can be expressed as $v_1\mathbf{i} + v_2\mathbf{j}$. This is a **linear combination** of $\mathbf{i}$ and $\mathbf{j}$.

Example 6   Express the vector $\mathbf{r} = \langle 2, -6 \rangle$ as a linear combination of $\mathbf{i}$ and $\mathbf{j}$.

$$\mathbf{r} = \langle 2, -6 \rangle = \boxed{}\mathbf{i} + (-6)\mathbf{j} = 2\mathbf{i} - \boxed{}\mathbf{j}$$

Example 7   Write the vector $\mathbf{q} = -\mathbf{i} + 7\mathbf{j}$ in component form.

$$\mathbf{q} = -\mathbf{i} + 7\mathbf{j} = \boxed{}\mathbf{i} + \boxed{}\mathbf{j} = \langle -1, \boxed{} \rangle$$

Example 8   If $\mathbf{a} = 5\mathbf{i} - 2\mathbf{j}$ and $\mathbf{b} = -\mathbf{i} + 8\mathbf{j}$, find $3\mathbf{a} - \mathbf{b}$.

$$3\mathbf{a} - \mathbf{b} = \boxed{}(5\mathbf{i} - 2\mathbf{j}) - (\boxed{} + 8\mathbf{j})$$
$$= 15\mathbf{i} - \boxed{}\mathbf{j} + \mathbf{i} - \boxed{}\mathbf{j}$$
$$= \boxed{}\mathbf{i} - 14\mathbf{j}$$

Example 9   Calculate and sketch the unit vector $\mathbf{u} = (\cos\theta)\mathbf{i} + (\sin\theta)\mathbf{j}$ for $\theta = 2\pi/3$. Include the unit circle in your sketch.

$$\mathbf{u} = \left(\cos\frac{2\pi}{3}\right)\mathbf{i} + \left(\sin\frac{2\pi}{3}\right)\mathbf{j} = \boxed{}\mathbf{i} + \boxed{}\mathbf{j}$$

Now sketch this vector.

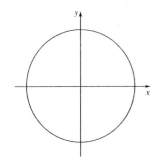

Example 10 Determine the direction angle θ of the vector $\mathbf{w} = -4\mathbf{i} - 3\mathbf{j}$.

$\mathbf{w} = \langle -4, -3 \rangle$

$\tan\theta = \dfrac{-3}{\boxed{}} = \dfrac{3}{4}$, so $\tan^{-1}\dfrac{3}{4} = \boxed{}$.

Since $\mathbf{w}$ is in quadrant $\boxed{}$, θ is a third-quadrant angle with reference angle $\tan^{-1}\dfrac{3}{4} \approx 37°$. Thus, $\theta \approx \boxed{}° + 37°$, or $217°$.

Example 11 *Airplane Speed and Direction.* An airplane travels on a bearing of $100°$ at an airspeed of 190 km/h while a wind is blowing 48 km/h from $220°$. Find the ground speed of the airplane and the direction of its track, or course, over the ground.

We make a drawing.

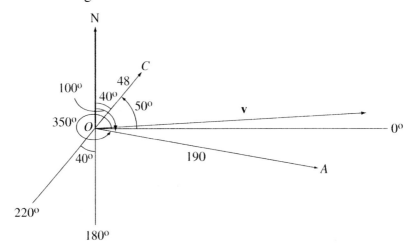

$\mathbf{v} = \overline{OA} + \overline{OC}$

$= \left[190(\cos 350°)\mathbf{i} + 190(\sin 350°)\mathbf{j}\right] + \left[48(\cos 50°)\mathbf{i} + 48(\sin 50°)\mathbf{j}\right]$

$= \left[190(\cos 350°) + \boxed{}(\cos 50°)\right]\mathbf{i} + \left[\boxed{}(\sin 350°) + 48(\sin 50°)\right]\mathbf{j}$

$\approx 217.97\mathbf{i} + 3.78\mathbf{j}$

Ground speed $= |\mathbf{v}| \approx \sqrt{(217.97)^2 + \left(\boxed{}\right)^2} \approx 218$ km/h

Now let α be the direction angle of $\mathbf{v}$.

$\tan\alpha \approx \dfrac{3.78}{\boxed{}}$

$\alpha = \tan^{-1}\dfrac{3.78}{217.97} \approx \boxed{}°$

The course of the airplane (its direction from north) is $90° - 1°$, or $\boxed{}°$.

Dot Product

The **dot product** of two vectors $\mathbf{u} = \langle u_1, u_2 \rangle$ and $\mathbf{v} = \langle v_1, v_2 \rangle$ is

$$\mathbf{u} \cdot \mathbf{v} = u_1 v_1 + u_2 v_2.$$

(Note that $u_1 v_1 + u_2 v_2$ is a *scalar*, not a vector.)

Example 12   Find the indicated dot product when

$$\mathbf{u} = \langle 2, -5 \rangle, \ \mathbf{v} = \langle 0, 4 \rangle, \ \text{and} \ \mathbf{w} = \langle -3, 1 \rangle.$$

a) $\mathbf{u} \cdot \mathbf{w}$ b) $\mathbf{w} \cdot \mathbf{v}$

a) $\mathbf{u} \cdot \mathbf{w} = 2(-3) + (-5)(\boxed{})$

$\qquad = \boxed{} - 5 = \boxed{}$

b)   Note that $\mathbf{w} \cdot \mathbf{v} = \mathbf{v} \cdot \mathbf{w}$.

$\qquad \mathbf{w} \cdot \mathbf{v} = \boxed{}(-3) + 4 \cdot \boxed{} = \boxed{}$

Angle Between Two Vectors

If θ is the angle between two *nonzero* vectors $\mathbf{u}$ and $\mathbf{v}$, then

$$\cos \theta = \frac{\mathbf{u} \cdot \mathbf{v}}{|\mathbf{u}||\mathbf{v}|}.$$

Example 13   Find the angle between $\mathbf{u} = \langle 3, 7 \rangle$ and $\mathbf{v} = \langle -4, 2 \rangle$.

$$\mathbf{u} \cdot \mathbf{v} = \boxed{}(-4) + 7 \cdot \boxed{} = -12 + 14 = \boxed{}$$

$$|\mathbf{u}| = \sqrt{3^2 + 7^2} = \sqrt{\boxed{}}$$

$$|\mathbf{v}| = \sqrt{(-4)^2 + \boxed{}^2} = \sqrt{20}$$

$$\cos \alpha = \frac{\mathbf{u} \cdot \mathbf{v}}{|\mathbf{u}||\mathbf{v}|} = \frac{\boxed{}}{\sqrt{58}\sqrt{20}}$$

$$\theta = \cos^{-1} \frac{2}{\sqrt{58}\sqrt{20}} \approx \boxed{}^{\circ}$$

Example 14 A 350-lb block is suspended by two cables, as shown below. At point A, there are three forces acting: **W**, the block pulling down, and **R** and **S**, the two cables pulling upward and outward, respectively. Find the tension in each cable.

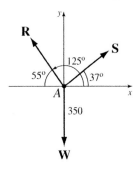

$$\mathbf{R} = |\mathbf{R}|\left[\left(\cos 125°\right)\mathbf{i} + \left(\sin 125°\right)\mathbf{j}\right]$$

$$\mathbf{S} = |\mathbf{S}|\left[\left(\cos 37°\right)\mathbf{i} + \left(\sin 37°\right)\mathbf{j}\right]$$

$$\mathbf{W} = |\mathbf{W}|\left[\left(\cos 270°\right)\mathbf{i} + \left(\sin 270°\right)\mathbf{j}\right] = \boxed{}\mathbf{j}$$

The forces are acting in equilibrium, so their sum is equal to the zero vector $\mathbf{0} = 0\mathbf{i} + 0\mathbf{j}$.

$$\mathbf{R} + \mathbf{S} + \mathbf{W} = \mathbf{0}$$

$$|\mathbf{R}|\left[\left(\cos 125°\right)\mathbf{i} + \left(\sin 125°\right)\mathbf{j}\right] + |\mathbf{S}|\left[\left(\cos 37°\right)\mathbf{i} + \left(\sin 37°\right)\mathbf{j}\right] - 350\mathbf{j} = 0\boxed{} + \boxed{}\mathbf{j}$$

This give us two equations.

$$|\mathbf{R}|\cos 125° + |\mathbf{S}|\cos 37° = 0,$$

$$|\mathbf{R}|\sin 125° + |\mathbf{S}|\sin 37° - 350 = 0$$

Solving the first equation for $|\mathbf{R}|$, we get

$$|\mathbf{R}| = -\frac{|\mathbf{S}|\left(\cos 37°\right)}{\cos 125°}$$

Substituting this expression for $|\mathbf{R}|$ in the second equation and solving for $|\mathbf{S}|$, we get $|\mathbf{S}| \approx 201$ lb. Then substituting in the expression we found for $|\mathbf{R}|$, we have $|\mathbf{R}| \approx 280$ lb.

The tension in the cable on the left is $\boxed{}$ lb, and the tension in the cable on the right is $\boxed{}$ lb.

Section 9.1   Systems of Equations in Two Variables

A **system of equations** is composed of two or more equations considered simultaneously. For example,

$$x + y = 6,$$
$$3x - y = 2$$

is a **system of two linear equations in two variables**. The solution set of this system consists of all ordered pairs that make *both* equations true.

When we graph a system of linear equations, each point at which the graphs intersect is a **solution of the system of equations**.

Example 1   Solve the system of equations graphically.

$$x - y = 5,$$
$$2x + y = 1$$

We graph the equations on the same set of axes.

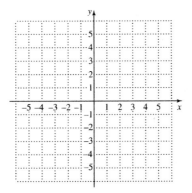

The graphs intersect at $\left(\boxed{}, -3 \right)$, so $\left(\boxed{}, -3 \right)$ is the solution of this system of equations.

One algebraic technique for solving systems of equations is the **substitution method**. In this method, we use one equation to express one variable in terms of the other. Then we substitute that expression in the other equation.

Example 2   Use the substitution method to solve the system

$$x - y = 5, \qquad (1)$$
$$2x + y = 1. \qquad (2)$$

First, we solve equation (1) for x.

$$x = \boxed{}$$

Now, substitute $y + 5$ for x in equation (2) and solve for y.

$$2x + y = 1$$

$$2\left(\boxed{}\right) + y = 1$$

$$2y + 10 + y = 1$$

$$\boxed{} = 1 \qquad \text{Collecting like terms}$$

$$3y = \boxed{} \qquad \text{Subtracting 10}$$

$$y = \boxed{} \qquad \text{Dividing by 3}$$

Now, substitute -3 for y into either of the original equations and solve for x. We use equation (1).

$$x - y = 5$$
$$x - (-3) = 5$$
$$x + \boxed{} = 5$$
$$x = \boxed{}$$

We check $(2, -3)$ in both equations.

$x - y = 5$		$2x + y = 1$	
$2 - (-3) \overset{?}{\,} 5$		$2 \cdot 2 + (-3) \overset{?}{\,} 1$	
$2 + \boxed{}$		$4 - \boxed{}$	
$\boxed{}$	5 TRUE	$\boxed{}$	1 TRUE

The solution is $(2, -3)$.

Another algebraic method for solving systems of equations is the **elimination method**. In this method we eliminate a variable by adding two equations.

Example 3 Use the elimination method to solve the system of equations.
$$2x + y = 2, \qquad (1)$$
$$x - y = 7. \qquad (2)$$

The coefficient of y in equation (1) is 1, and in equation (2) it is -1. Adding the two equations gives

$$
\begin{array}{rcl}
2x + y & = & 2 \\
x - y & = & 7 \\
\hline
3x & = & \boxed{} \qquad \text{Adding} \\
x & = & \boxed{} \qquad \text{Dividing by 3}
\end{array}
$$

Substitute 3 for x in either of the original equations (This is called back-substitution.) and solve
for y. We use equation (1).

$$2x + y = 2$$
$$2 \cdot \boxed{} + y = 2$$
$$6 + y = 2$$
$$y = \boxed{}$$

A check shows that the ordered pair $(3, -4)$ is a solution of both equations, so the solution is
$\left(\boxed{}, -4 \right)$.

Sometimes it might be necessary to multiply one or both equations by constants in order to find two equations in which the coefficients of a variable are opposites.

Example 4 Use the elimination method to solve the system of equations
$$4x + 3y = 11, \qquad (1)$$
$$-5x + 2y = 15. \qquad (2)$$

We observe that if we multiply the first equation by 5 and the second equation by 4, the coefficients of the x-terms will be opposites. This will allow us to eliminate x.

$$
\begin{array}{rcl}
20x + \boxed{} & = & 55 \\
\boxed{} + \quad 8y & = & 60 \\
\hline
23y & = & \boxed{} \\
y & = & 5
\end{array}
$$

We can substitute 5 for y in either of the original equations and solve for x. We use the first equation.

$$4x + 3 \cdot \boxed{} = 11$$

$$4x + \boxed{} = 11$$

$$4x = \boxed{}$$

$$x = -1$$

The pair $(-1, 5)$ checks, so it is the solution.

Example 5   Solve each of the following systems using the elimination method

a) $x - 3y = 1$,   (1)

$-2x + 6y = 5$ (2)

Multiply equation (1) by 2 and then add.

$$2x - 6y = \boxed{}$$

$$\underline{-2x + 6y = 5}$$

$$0 = \boxed{} \qquad \text{False}$$

Because we get a false equation, the system of equations has no solution. The solution set is $\varnothing$. The system of equations is inconsistent, and the equations are _____.
               independent / dependent

b) $2x + 3y = 6$,   (1)

$4x + 6y = 12$ (2)

Multiply equation (1) by -2 and then add.

$$-4x - 6y = \boxed{}$$

$$\underline{4x + 6y = 12}$$

$$0 = \boxed{} \qquad \text{True}$$

We obtain an equation that is true for all values of x and y. This tells us that the equations are dependent, so the system of equations has _____ solutions.
                                        no / infinitely many

We can find these solutions by solving one of the equations for x or for y. Let's solve equation (1) for y.

$$2x + 3y = 6$$

$$3y = -2x + \boxed{}$$

$$y = -\frac{2}{3}x + \boxed{}$$

We can express the solutions as $\left(x, -\dfrac{2}{3}x+2 \right)$.

If $x = -3$: $-\dfrac{2}{3}(-3)+2 = 2+2 = \boxed{}$.

If $x = 0$: $-\dfrac{2}{3}\cdot 0+2 = \boxed{}$.

If $x = 6$: $-\dfrac{2}{3}\cdot 6+2 = -4+2 = \boxed{}$.

So three solutions of the system of equations are

$(-3, 4)$, $\left(0, \boxed{}\right)$, and $\left(\boxed{}, -2\right)$.

We could also solve one of the equations for x. For example, if we solve $2x + 3y = 6$ for

x, we have $x = -\dfrac{3}{2}y+3$. Then the solutions can be expressed as $\left(-\dfrac{3}{2}y+3,\ y \right)$.

Example 6 *Snack Mixtures.* At Max's Munchies, caramel corn worth $2.50 per pound is mixed with honey roasted mixed nuts worth $7.50 per pound in order to get 20 lb of a mixture worth $4.50 per pound. How much of each snack is used?

1. **Familiarize.**

   To begin with, guess a possible solution: 16 lb caramel corn, 4 lb nuts

   Total weight: $\boxed{}$ lb caramel corn $+ \boxed{}$ lb nuts $= 20$ lb.

   Total value: Caramel corn: $2.50(16) = \boxed{}$

   Nuts: $7.50(4) = \underline{\boxed{}}$

   $\boxed{}$

   A 20-lb mixture price at $4.50 per pound would have a value of $4.50(20) = \boxed{}$. Our guess is not correct, but it helps us translate to an equation.

2. **Translate.** Let $x =$ the number of pounds of caramel corn in the mixture and $y =$ the number of pounds of nuts. We organize the information in a table.

	Caramel Corn	Nuts	Mixture
Price per Pound	$2.50	$\boxed{}$	$4.50
Number of Pounds	x	y	20
Value of Mixture	$2.50x$	$7.50y$	$4.50(20)$, or $\boxed{}$

The last two rows of the table give us two equations.

$$x + y = 20$$
$$2.50x + 7.50y = 90, \text{ or } 2.5x + 7.5y = 90$$

We multiply the second equation by 10 to clear the decimals and we have the system

$$x + y = 20,$$
$$25x + 75y = 900.$$

3. **Carry out**. We multiply the first equation by -25 and then add.

$$-25x - 25y = \boxed{}$$
$$\underline{25x + 75y = 900}$$
$$50y = \boxed{}$$
$$y = \boxed{}$$

Now substitute in one of the original equations to find x. We use the first equation.

$$x + \boxed{} = 20$$
$$x = \boxed{}$$

4. **Check**.   12 lb of caramel corn

   8 lb of nuts

   $12 \text{ lb} + 8 \text{ lb} = \boxed{}$ lb, so the weight checks.

   $\$2.50(12) + \$7.50(8) = \$30 + \boxed{} = \boxed{}$, so the value of the mixture also checks.

5. **State**. The mixture should consist of $\boxed{}$ lb of caramel corn and $\boxed{}$ lb of honey roasted mixed nuts.

Example 7   *Airplane Travel*. An airplane flies the 3000-mi distance from Los Angeles to New York, with a tailwind, in 5 hr. The return trip, against the wind, takes 6 hr. Find the speed of the airplane in still air and the speed of the wind.

1. **Familiarize**. Let $p =$ the speed of the plane and $w =$ the speed of the wind.

   Distance of each trip: $\boxed{}$ mi

   Speed with a tailwind: $p + w$

   Time with a tailwind: $\boxed{}$ hr

   Speed with a headwind: $p - w$

   Time with a headwind: $\boxed{}$ hr

2. **Translate**.

	Distance	Rate	Time
With Tailwind	3000	$p+w$	
With Headwind	3000	$p-w$	

Using Distance $=$ Rate $\times$ Time, the table gives us two equations.

$3000 = (p+w)5$, or $600 = p+w$ (1)

$3000 = (p-w)6$, or $500 = p-w$ (2)

3. **Carry out**.

$600 = p+w$ (1)

$500 = p-w$ (2)

$\boxed{} = 2p$ Adding

$\boxed{} = p$

Use either equation to find w. We use equation (1)

$600 = \boxed{} + w$

$\boxed{} = w$

4. **Check**.

Speed with tailwind: $550 + 50 = 600$ mph

Speed with headwind: $550 - \boxed{} = 500$ mph

Time with tailwind: $t = \dfrac{d}{r} = \dfrac{3000}{600} = \boxed{}$ hr

Time with headwind: $t = \dfrac{d}{r} = \dfrac{3000}{500} = \boxed{}$ hr

The answer checks.

5. **State**. The speed of the plane is 550 mph, and the speed of the wind is $\boxed{}$ mph.

Example 8   *Supply and Demand.* Suppose that the price and supply of the Star Station satellite radio are related by the equation

$$y = 90 + 30x,$$

where y is the price, in dollars, at which the seller is willing to supply x thousand units. Also suppose that the price and the demand for the same model of satellite radio are related by the equation

$$y = 200 - 25x,$$

where y is the price, in dollars, at which the consumer is willing to buy x thousand units. The **equilibrium point** for this radio is the pair (x, y) that is a solution of both equations. The **equilibrium price** is the price at which the amount of product that the seller is willing to supply is the same as the amount demanded by the customer. Find the equilibrium point for this radio.

1., 2.   Familiarize and Translate.

$$y = 90 + 30x,$$
$$y = 200 - 25x$$

We substitute some values of x in these equations to familiarize ourselves with the problem.

Let $x = 1$: $\quad y = 90 + 30 \cdot \boxed{} = 120$

$\qquad\qquad\quad y = \boxed{} - 25 \cdot 1 = 175$

The value of y related to supply is less than the value related to demand.

Let $x = 4$: $\quad y = 90 + 30 \cdot 4 = \boxed{}$

$\qquad\qquad\quad y = 200 - 25 \cdot \boxed{} = 100$

In this case the value of y related to supply is higher than the value related to demand. We see that the x-value we are looking for is between 1 and 4.

3. Carry out. We use the substitution method.

$$90 + 30x = 200 - 25x$$

$90 + 55x = \boxed{} \qquad$ Adding $25x$

$55x = \boxed{} \qquad$ Subtracting 90

$x = 2$

4. Check. When we substitute 2 for x and 150 for y in the two equations we were given, we find that the answer checks.

5. State. The equilibrium point is $\left(\boxed{}, \$150 \right)$. That is, the equilibrium quantity is 2 thousand units and the equilibrium price is $\boxed{}$.

Section 9.2   Systems of Equations in Three Variables

A solution of a system of three linear equations in three variables is an ordered triple that is a solution of all three equations.

We can use an algebraic method called **Gaussian elimination** to solve systems of equations in three variables. The goal is to transform the system of equations into a system of the form

$$Ax + By + Cz = D,$$
$$Ey + Fz = G,$$
$$Hz = K.$$

Once z has been found, back-substitution can be used to find y and then x. To transform the original system to the desired form, any of the following operations can be applied.

1. Interchange any two equations.
2. Multiply both sides of one of the equations by a nonzero constant.
3. Add a nonzero multiple of one equation to another equation.

Example 1   Solve the following system:

$$x - 2y + 3z = 11, \quad (1)$$
$$4x + 2y - 3z = 4, \quad (2)$$
$$3x + 3y - z = 4. \quad (3)$$

First we multiply equation (1) by -4 and add it to equation (2) to eliminate x.

$$
\begin{array}{r}
-4x + 8y - 12z = -44 \\
4x + 2y - 3z = 4 \\
\hline
\boxed{} - 15z = -40 \quad (4)
\end{array}
$$

Next multiply equation (1) by -3 and add it to equation (3) to eliminate x.

$$
\begin{array}{r}
-3x + 6y - 9z = -33 \\
3x + 3y - z = 4 \\
\hline
9y - 10z = \boxed{} \quad (5)
\end{array}
$$

This gives us

$$x - 2y + 3z = 11, \quad (1)$$
$$10y - 15z = -40, \quad (4)$$
$$9y - 10z = -29. \quad (5)$$

Now multiply equation (5) by 10 to make the y-coefficient a multiple of the y-coefficient in equation (4).

$$x - 2y + 3z = 11, \quad (1)$$
$$10y - 15z = -40, \quad (4)$$
$$\boxed{} - 100z = -290. \quad (6)$$

Multiply equation (4) by -9 and add it to equation (6).

$$-90y + 135z = 360$$
$$\underline{90y - 100z = -290 \qquad (6)}$$
$$35z = \boxed{} \qquad (7)$$

This gives us

$$x - 2y + 3z = 11, \quad (1)$$
$$10y - 15z = -40, \quad (4)$$
$$35z = 70. \quad (7)$$

Solve equation (7) for z.

$$35z = 70$$
$$z = \boxed{}$$

Now substitute 2 for z in equation (4) and solve for y.

$$10y - 15z = -40$$
$$10y - 15 \cdot 2 = -40$$
$$10y - \boxed{} = -40$$
$$10y = -10$$
$$y = \boxed{}$$

Substitute 2 for z and -1 for y in equation (1) and solve for x.

$$x - 2y + 3z = 11$$
$$x - 2(-1) + 3(2) = 11$$
$$x + 2 + 6 = 11$$
$$x = \boxed{}$$

The solution is $(3, -1, 2)$.

Example 2   Solve the following system:

$$x + y + z = 7, \quad (1)$$
$$3x - 2y + z = 3, \quad (2)$$
$$x + 6y + 3z = 25. \quad (3)$$

Begin by eliminating x from equations (2) and (3). We multiply equation (1) by -3 and add it to equation (2). We also multiply equation (1) by -1 and add it to equation (3).

$$x + y + z = 7 \quad (1)$$
$$-5y - 2z = \boxed{} \quad (4)$$
$$\boxed{} + 2z = 18 \quad (5)$$

Next add equations (4) and (5) to eliminate y in equation (5).

$$x + y + z = 7$$
$$-5y - 2z = -18$$
$$0 = \boxed{}$$

The equation $0 = 0$ tells us that equations (1), (2), and (3) are dependent. In this case there are _____ solutions.
  no / infinitely many

It can be shown that equation (3) is dependent on equations (1) and (2), so the original system is equivalent to

$$x + y + z = 7, \quad (1)$$
$$3x - 2y + z = 3. \quad (2)$$

We can express the solutions in terms of one of the variables. We begin by solving equation (4) for y.

$$-5y - 2z = -18$$
$$-5y = \boxed{} - 18$$
$$y = -\frac{2}{5}z + \boxed{} \qquad \text{Dividing by } -5$$

Substitute $-\dfrac{2}{5}z + \dfrac{18}{5}$ for y in equation (1) and solve for x.

$$x - \frac{2}{5}z + \frac{18}{5} + z = 7$$
$$x = -\frac{3}{5}z + \frac{17}{5}$$

Then the solutions are $\left(-\dfrac{3}{5}z + \dfrac{17}{5}, \ -\dfrac{2}{5}z + \dfrac{18}{5}, \ \boxed{} \right).$

Example 3   Luis inherited $15,000 and invested part of it in a money market account, part in municipal bonds, and part in a mutual fund. After 1 yr, he received a total of $730 in simple interest from the three investments. The money market account paid 4% annually, the bonds paid 5% annually, and the mutual fund paid 6% annually. There was $2000 more invested in the mutual fund than in bonds. Find the amount that Luis invested in each category.

1. **Familiarize.** Let x, y, and z represent the amounts invested in the money market account, the bonds, and the mutual fund, respectively. The income produced by the investments is given by $4\% \cdot x$, $\boxed{}\% \cdot y$, and $6\% \cdot z$, or $0.04x$, $0.05y$, and $0.06z$.

2. **Translate.** A total of $\boxed{}$ is invested, so

    $$x + y + z = \boxed{}.$$

    The total interest is $730, so

    $$0.04x + \boxed{} + 0.06z = 730.$$

    The amount invested in the mutual fund was $2000 more than the amount invested in bonds, so we have

    $$z = 2000 + y.$$

    We have a system of equations. We multiply the second equation by 100 to clear decimals.

    $$x + y + z = \boxed{}$$
    $$4x + 5y + 6z = 73,000$$
    $$z = 2000 + y$$

3. **Carry out.** The solution is $(7000, 3000, 5000)$.

4. **Check.** The sum of the number is 15,000. The interest is
    $$0.04(7000) + 0.05(3000) + 0.06(5000) = 280 + \boxed{} + 300 = \boxed{}.$$ The amount invested in the mutual fund, $5000, is $2000 more than the amount invested in bonds, $3000. The answer checks.

5. **State.** Luis invested $\boxed{}$ in a money market account, $\boxed{}$ in bonds, and $\boxed{}$ in a mutual fund.

Example 4   The following table lists the number of civil cases pending in U.S. federal courts in three recent years. Use the data to find a quadratic function that gives the number of pending civil cases as a function of the number of years after 2009. Then use this function to estimate the number of civil cases pending in U.S. federal courts in 2014.

Year, x	Number of Civil Cases Pending in U.S. Federal Courts
2009, 0	311,353
2011, 2	267,495
2013, 4	300,469

Let $x =$ the number of years after 2009. We have three data points: $(0,\ 311{,}353)$, $\left(\boxed{},\ 267{,}495\right)$, and $(4,\ 300{,}469)$. Substitute in $n(x) = ax^2 + bx + c$.

$$311{,}353 = a \cdot 0^2 + b \cdot 0 + c$$
$$311{,}353 = c$$

$$267{,}495 = a \cdot 2^2 + b \cdot 2 + c$$
$$267{,}495 = 4a + \boxed{}\,b + c$$

$$300{,}469 = a \cdot 4^2 + b \cdot \boxed{} + c$$
$$300{,}469 = \boxed{}\,a + 4b + c$$

We have a system of equations:

$$311{,}353 = c,$$
$$267{,}495 = 4a + 2b + c,$$
$$\boxed{} = 16a + 4b + c.$$

Substitute 311,353 for c in the last two equations and simplify.

$$-43{,}858 = 4a + \boxed{}\,b,$$
$$-10{,}884 = 16a + 4b$$

Divide the first equation by 2, and divide the second equation by 4.

$$-21{,}929 = 2a + b$$
$$-2{,}721 = 4a + \boxed{}$$

Multiply the first equation by -1 and then add the equations.

$$21,929 = -2a - b$$
$$-2,721 = \;\;4a + b$$
$$\overline{19,208 = \boxed{}}$$
$$9604 = a$$

Now substitute and solve for b.

$$2a + b = -21,929$$
$$19,208 + b = -21,929$$
$$b = \boxed{}$$

$$n(x) = 9604x^2 - 41,137x + \boxed{}$$

In 2014, $x = \boxed{}$.

$$n(5) = 9604 \cdot 5^2 - 41,137 \cdot 5 + 311,353$$
$$\approx 345,768$$

We estimate that there were $\boxed{}$ civil cases pending in U.S. federal courts in 2014.

Section 9.3   Matrices and Systems of Equations

We can use matrices and Gaussian elimination to solve systems of equations. A **matrix** is a rectangular array of numbers. We can use the following operations to transform a matrix like

$$\begin{bmatrix} 2 & -1 & 4 & -3 \\ 1 & -2 & -10 & -6 \\ 3 & 0 & 4 & 7 \end{bmatrix}$$

to a row-equivalent matrix of the form

$$\begin{bmatrix} 1 & a & b & c \\ 0 & 1 & d & e \\ 0 & 0 & 1 & f \end{bmatrix}.$$

Row-Equivalent Operations

1. Interchange any two rows.
2. Multiply each entry in a row by the same nonzero constant.
3. Add a nonzero multiple of one row to another row.

Example 1   Solve the following system:

$$2x - y + 4z = -3,$$
$$x - 2y - 10z = -6,$$
$$3x \qquad + 4z = 7.$$

First we write the augmented matrix for this system of equations. This matrix consists of the coefficients of the x-, y-, and z-terms, along with the constant terms. We write a 0 for the missing y-term in the third equation.

$$\begin{bmatrix} 2 & -1 & 4 & \Box \\ 1 & -2 & \Box & -6 \\ 3 & \Box & 4 & 7 \end{bmatrix}$$

Interchanging rows 1 and 2 puts a 1 in the upper left-hand corner, and will help to simplify the process of putting the first column in the proper form.

$$\begin{bmatrix} 1 & \Box & -10 & -6 \\ \Box & -1 & 4 & -3 \\ 3 & 0 & 4 & \Box \end{bmatrix} \begin{matrix} \text{New row 1 = row 2} \\ \text{New row 2 = row 1} \\ \\ \end{matrix}$$

Now we multiply the first row by -2 and add it to row 2, and then multiply the first row by -3 and add it to row 3.

$$\begin{bmatrix} 1 & -2 & -10 & | & -6 \\ \square & 3 & 24 & | & \square \\ 0 & \square & 34 & | & \square \end{bmatrix} \quad \begin{array}{l} \text{New row } 2 = -2(\text{row } 1) + \text{row } 2 \\ \text{New row } 3 = -3(\text{row } 1) + \text{row } 3 \end{array}$$

To get a 1 in the second row, second column, multiply row 2 by $\dfrac{1}{3}$.

$$\begin{bmatrix} 1 & -2 & -10 & | & -6 \\ 0 & \square & 8 & | & 3 \\ \square & 6 & 34 & | & \square \end{bmatrix} \quad \text{New row } 2 = \frac{1}{3}(\text{row } 2)$$

We can get a 0 in the third row, second column by multiplying the second row by -6 and adding it to row 3.

$$\begin{bmatrix} \square & -2 & -10 & | & \square \\ 0 & \square & 8 & | & 3 \\ 0 & \square & -14 & | & 7 \end{bmatrix} \quad \text{New row } 3 = -6(\text{row } 2) + \text{row } 3$$

Finally, we multiply the last row of the matrix by $-\dfrac{1}{14}$ to get a 1 in the third row, third column.

$$\begin{bmatrix} 1 & -2 & -10 & | & \square \\ 0 & \square & 8 & | & 3 \\ \square & 0 & \square & | & -\frac{1}{2} \end{bmatrix} \quad \text{New row } 3 = -\frac{1}{14}(\text{row } 3)$$

Now we reinsert the variables and write the system of equations that corresponds to the last matrix.

$$x - 2y - 10z = -6, \qquad (1)$$

$$\boxed{} + 8z = 3, \qquad (2)$$

$$z = \boxed{} \qquad (3)$$

Substitute $-\dfrac{1}{2}$ for z in equation (2) and solve for y.

$$y + 8\left(-\dfrac{1}{2}\right) = 3$$

$$y - 4 = 3$$

$$y = \boxed{}$$

Now substitute for y and z in equation (1) and solve for x.

$$x - 2 \cdot \boxed{} - 10\left(-\dfrac{1}{2}\right) = -6$$

$$\boxed{} - 14 + 5 = -6$$

$$x = \boxed{}$$

The solution of the system of equations is $\left(3,\, 7,\, -\dfrac{1}{2}\right)$.

The goal of Gaussian elimination is to find a matrix in the following form:

$$\left[\begin{array}{ccc|c} 1 & a & b & c \\ 0 & 1 & d & e \\ 0 & 0 & 1 & f \end{array}\right].$$

This is called **row-echelon form**.

Row-Echelon Form

1. If a row does not consist entirely of 0's, then the first nonzero element in the row is a 1 (called a **leading 1**).

2. For any two successive nonzero rows, the leading 1 in the lower row is farther to the right than the leading 1 in the higher row.

3. All the rows consisting entirely of 0's are at the bottom of the matrix.

If a fourth property is also satisfied, a matrix is said to be in **reduced row-echelon form**.

4. Each column that contains a leading 1 has 0's everywhere else.

Example 2 Which of the following matrices are in row-echelon form? Which, if any, are in reduced row-echelon form?

a) $\begin{bmatrix} 1 & -3 & 5 & | & -2 \\ 0 & 1 & -4 & | & 3 \\ 0 & 0 & 1 & | & 10 \end{bmatrix}$

b) $\begin{bmatrix} 0 & -1 & | & 2 \\ 0 & 1 & | & 5 \end{bmatrix}$

c) $\begin{bmatrix} 1 & -2 & -6 & 4 & | & 7 \\ 0 & 3 & 5 & -8 & | & -1 \\ 0 & 0 & 1 & 9 & | & 2 \end{bmatrix}$

d) $\begin{bmatrix} 1 & 0 & 0 & | & -2.4 \\ 0 & 1 & 0 & | & 0.8 \\ 0 & 0 & 1 & | & 5.6 \end{bmatrix}$

e) $\begin{bmatrix} 1 & 0 & 0 & 0 & | & 2/3 \\ 0 & 1 & 0 & 0 & | & -1/4 \\ 0 & 0 & 1 & 0 & | & 6/7 \\ 0 & 0 & 0 & 1 & | & 0 \end{bmatrix}$

f) $\begin{bmatrix} 1 & -4 & 2 & | & 5 \\ 0 & 0 & 0 & | & 0 \\ 0 & 1 & -3 & | & 8 \end{bmatrix}$

a) This matrix _____ in row-echelon form. It _____ in reduced
   is / is not   is / is not
   row-echelon form.

b) This matrix _____ in row-echelon form.
   is / is not

c) This matrix _____ in row-echelon form.
   is / is not

d) This matrix _____ in row-echelon form. It _____ in reduced
   is / is not   is / is not
   row-echelon form.

e) This matrix _____ in row-echelon form. It _____ in reduced
   is / is not   is / is not
   row-echelon form.

f) This matrix _____ in row-echelon form.
   is / is not

> The goal of **Gauss-Jordan elimination** is to find a *reduced* row-echelon matrix that is equivalent to a given system of equations.

Example 3 Use Gauss-Jordan elimination to solve the system of equations

$$2x - y + 4z = -3, \quad (1)$$
$$x - 2y - 10z = -6 \quad (2)$$
$$3x \quad + 4z = 7. \quad (3)$$

Earlier we solved this system using Gaussian elimination and obtain the matrix

$$\begin{bmatrix} 1 & -2 & -10 & | & -6 \\ 0 & 1 & 8 & | & 3 \\ 0 & 0 & 1 & | & -1/2 \end{bmatrix}.$$

We continue to perform row-equivalent operations until we have a matrix in reduced row-echelon form.

$$\begin{bmatrix} 1 & -2 & \square & -11 \\ 0 & 1 & \square & 7 \\ 0 & 0 & \square & -1/2 \end{bmatrix}$$ New row $1 = 10(\text{row } 3) + \text{row } 1$
New row $2 = -8(\text{row } 3) + \text{row } 2$

$$\begin{bmatrix} 1 & \square & 0 & \square \\ 0 & \square & 0 & \square \\ \square & 0 & 1 & \square \end{bmatrix}$$ New row $1 = 2(\text{row } 2) + \text{row } 1$

The matrix is now in reduced row-echelon form. The system of equations that corresponds to the matrix is

$$x = 3,$$
$$y = \square,$$
$$z = -\frac{1}{2}.$$

The solution is $\left(3, 7, -\dfrac{1}{2}\right)$.

Example 4   Solve the following system:

$$3x - 4y - z = 6 \quad (1)$$
$$2x - y + z = -1, \quad (2)$$
$$4x - 7y - 3z = 13. \quad (3)$$

First we write the augmented matrix. We will use Gauss-Jordan elimination.

$$\begin{bmatrix} 3 & -4 & \square & 6 \\ 2 & -1 & \square & -1 \\ \square & -7 & -3 & \square \end{bmatrix}$$

We then multiply the second and third rows by 3 so that each number in the first column below the first number, 3, is a multiple of that number.

$$\begin{bmatrix} 3 & -4 & -1 & | & 6 \\ 6 & \Box & 3 & \Box \\ 12 & \Box & \Box & | & 39 \end{bmatrix} \quad \begin{matrix} \\ \text{New row } 2 = 3(\text{row } 2) \\ \text{New row } 3 = 3(\text{row } 3) \end{matrix}$$

Now to get zeroes below the 3 in the first column, we can multiply the first row by -2 and add it to the second row and multiply the first row by -4 and add it to the third row.

$$\begin{bmatrix} 3 & -4 & \Box & | & 6 \\ \Box & 5 & 5 & | & -15 \\ \Box & -5 & \Box & | & 15 \end{bmatrix} \quad \begin{matrix} \\ \text{New row } 2 = -2(\text{row } 1) + \text{row } 2 \\ \text{New row } 3 = -4(\text{row } 1) + \text{row } 3 \end{matrix}$$

To get a zero in the second column of the third row, we can add the second row to the third row.

$$\begin{bmatrix} 3 & \Box & -1 & | & 6 \\ 0 & 5 & \Box & | & -15 \\ 0 & \Box & 0 & \Box \end{bmatrix} \quad \begin{matrix} \\ \\ \text{New row } 3 = \text{row } 2 + \text{row } 3 \end{matrix}$$

We have a row of all zeros, so the equations are dependent, and the system is equivalent to the following system.

$$3x - 4y - z = 6,$$
$$5y + 5z = -15$$

This particular system has infinitely many solutions. (A system containing dependent equations could be inconsistent.) We solve the second equation for y.

$$5y + 5z = -15$$
$$5y = \boxed{} - 15$$
$$y = -z - \boxed{}$$

Now we substitute $-z - 3$ for y in the first equation and solve for x.

$$3x - 4(-z - 3) - z = 6$$
$$3x + 4z + \boxed{} - z = 6$$
$$3x + 3z + 12 = 6$$
$$3x = -3z - \boxed{}$$
$$x = -z - \boxed{}$$

The solutions are of the form $\left(-z - 2, \ -z - 3, \ \boxed{} \right)$.

Section 9.4 Matrix Operations

We generally use a capital letter to represent a matrix and a lower-case letter with double subscripts to represent the entries. For example, a_{34} represents the entry which is in the third row and fourth column. The general term of a matrix is represented by a_{ij}, and we can write a matrix as

$$A = \begin{bmatrix} a_{ij} \end{bmatrix} = \begin{bmatrix} a_{11} & a_{12} & a_{13} & \cdots & a_{1n} \\ a_{21} & a_{22} & a_{23} & \cdots & a_{2n} \\ a_{31} & a_{32} & a_{33} & \cdots & a_{3n} \\ \vdots & \vdots & \vdots & & \vdots \\ a_{m1} & a_{m2} & a_{m3} & \cdots & a_{mn} \end{bmatrix}.$$

A matrix with m rows and n columns (as above) has order $m \times n$.

Addition and Subtraction of Matrices

Given two $m \times n$ matrices $A = \begin{bmatrix} a_{ij} \end{bmatrix}$ and $B = \begin{bmatrix} b_{ij} \end{bmatrix}$, their sum is

$$A + B = \begin{bmatrix} a_{ij} + b_{ij} \end{bmatrix}$$

and their difference is

$$A - B = \begin{bmatrix} a_{ij} - b_{ij} \end{bmatrix}.$$

Example 1 Find $A + B$ for each of the following.

a) $A = \begin{bmatrix} -5 & 0 \\ 4 & 1/2 \end{bmatrix}, \quad B = \begin{bmatrix} 6 & -3 \\ 2 & 3 \end{bmatrix}$

b) $A = \begin{bmatrix} 1 & 3 \\ -1 & 5 \\ 6 & 0 \end{bmatrix}, \quad B = \begin{bmatrix} -1 & -2 \\ 1 & -2 \\ -3 & 1 \end{bmatrix}$

a) A and B have the same order, 2×2, so we can add them.

$$A + B = \begin{bmatrix} -5+6 & 0+(-3) \\ 4+\square & 1/2+\square \end{bmatrix} = \begin{bmatrix} 1 & \square \\ 6 & \square \end{bmatrix}$$

b) A and B have the same order, 3×2, so we can add them.

$$A + B = \begin{bmatrix} 1+(-1) & \square+(-2) \\ -1+\square & 5+(-2) \\ \square+(-3) & 0+\square \end{bmatrix} = \begin{bmatrix} \square & 1 \\ 0 & \square \\ 3 & 1 \end{bmatrix}$$

Example 2   Find $C - D$ for each of the following.

a) $C = \begin{bmatrix} 1 & 2 \\ -2 & 0 \\ -3 & -1 \end{bmatrix}$, $D = \begin{bmatrix} 1 & -1 \\ 1 & 3 \\ 2 & 3 \end{bmatrix}$

b) $C = \begin{bmatrix} 5 & -6 \\ -3 & 4 \end{bmatrix}$, $D = \begin{bmatrix} -4 \\ 1 \end{bmatrix}$

a) C and D have the same order, 3×2, so we can subtract them.

$$C - D = \begin{bmatrix} 1 - \boxed{} & 2 - (-1) \\ -2 - \boxed{} & 0 - 3 \\ \boxed{} - 2 & -1 - \boxed{} \end{bmatrix} = \begin{bmatrix} \boxed{} & 3 \\ -3 & \boxed{} \\ -5 & \boxed{} \end{bmatrix}$$

b) C is a 2×2 matrix, and D is a 2×1 matrix. Since the order of C _____ the order
$$= / \neq$$

of D, we _____ subtract.
can / cannot

Example 3   Find $-A$ and $A + (-A)$ for

$$A = \begin{bmatrix} 1 & 0 & 2 \\ 3 & -1 & 5 \end{bmatrix}.$$

To find $-A$, we replace each entry of A by its opposite.

$$-A = \begin{bmatrix} -1 & \boxed{} & -2 \\ \boxed{} & 1 & \boxed{} \end{bmatrix}$$

$$A + (-A) = \begin{bmatrix} 1 & 0 & 2 \\ 3 & -1 & 5 \end{bmatrix} + \begin{bmatrix} -1 & 0 & -2 \\ -3 & 1 & -5 \end{bmatrix}$$

$$= \begin{bmatrix} 0 & \boxed{} & 0 \\ \boxed{} & 0 & \boxed{} \end{bmatrix}$$

Scalar Product

The **scalar product** of a number k and a matrix A is the matrix denoted by kA, obtained by multiplying each entry of A by the number k. The number k is called a **scalar**.

Example 4 Find $3A$ and $(-1)A$ for

$$A = \begin{bmatrix} -3 & 0 \\ 4 & 5 \end{bmatrix}.$$

$$3A = 3\begin{bmatrix} -3 & 0 \\ 4 & 5 \end{bmatrix} = \begin{bmatrix} \Box \cdot (-3) & 3 \cdot \Box \\ 3 \cdot \Box & \Box \cdot 5 \end{bmatrix} = \begin{bmatrix} -9 & \Box \\ 12 & \Box \end{bmatrix}$$

$$(-1)A = (-1)\begin{bmatrix} -3 & 0 \\ 4 & 5 \end{bmatrix} = \begin{bmatrix} 3 & \Box \\ \Box & -5 \end{bmatrix}$$

Example 5 Waterworks, Inc., manufactures three types of kayaks in its two plants. The table below lists the number of each style produced at each plant in April.

	Whitewater Kayak	Ocean Kayak	Crossover Kayak
Madison Plant	150	120	100
Greensburg Plant	180	90	130

a) Write a 2×3 matrix A that represents the information in the table.

b) The manufacturer increased production by 20% in May. Find a matrix M that represents the increased production figures.

c) Find the matrix $A + M$ and tell what it represents.

a) We record the information from the table in the matrix A below.

$$A = \begin{bmatrix} 150 & \Box & 100 \\ \Box & 90 & \Box \end{bmatrix}.$$

b) We write a matrix M that represents an increase of 20% in production.

$$M = A + 20\% \cdot A = A + 0.2A = \boxed{}A$$

$$= 1.2 \cdot \begin{bmatrix} 150 & 120 & 100 \\ 180 & 90 & 130 \end{bmatrix} = \begin{bmatrix} 180 & \Box & 120 \\ \Box & 108 & \Box \end{bmatrix}.$$

c) We find the sum $A + M$.

$$A + M = \begin{bmatrix} 330 & \Box & 220 \\ \Box & 198 & \Box \end{bmatrix}.$$

The sum $A + M$ represents the number of each type of kayak produced in each plant during $\Box$ and $\Box$.

Matrix Multiplication

For an $m \times n$ matrix $\mathbf{A} = \left[a_{ij} \right]$ and an $n \times p$ matrix $\mathbf{B} = \left[b_{ij} \right]$, the **product** $\mathbf{AB} = \left[c_{ij} \right]$ is an $m \times p$ matrix, where

$$c_{ij} = a_{i1} \cdot b_{1j} + a_{i2} \cdot b_{2j} + a_{i3} \cdot b_{3j} + \ldots + a_{in} \cdot b_{nj}$$

Example 6   For

$$\mathbf{A} = \begin{bmatrix} 3 & 1 & -1 \\ 2 & 0 & 3 \end{bmatrix}, \quad \mathbf{B} = \begin{bmatrix} 1 & 6 \\ 3 & -5 \\ -2 & 4 \end{bmatrix}, \quad \mathbf{C} = \begin{bmatrix} 4 & -6 \\ 1 & 2 \end{bmatrix},$$

find each of the following.

a) **AB**   b) **BA**   c) **BC**   d) **AC**

a) We first note that $\mathbf{A}$ has $\boxed{}$ columns, and $\mathbf{B}$ has $\boxed{}$ rows. Therefore, we

can perform this multiplication / cannot perform this multiplication .

$$\mathbf{AB} = \begin{bmatrix} 3 & 1 & -1 \\ 2 & 0 & 3 \end{bmatrix} \cdot \begin{bmatrix} 1 & 6 \\ 3 & -5 \\ -2 & 4 \end{bmatrix}$$

$$= \begin{bmatrix} 3 \cdot 1 + \boxed{} \cdot \boxed{} + (-1) \cdot (-2) & 3 \cdot 6 + \boxed{} \cdot (-5) + (-1) \cdot \boxed{} \\ 2 \cdot \boxed{} + 0 \cdot 3 + \boxed{} \cdot (-2) & \boxed{} \cdot 6 + 0 \cdot (-5) + \boxed{} \cdot 4 \end{bmatrix}$$

$$= \begin{bmatrix} 8 & \boxed{} \\ \boxed{} & 24 \end{bmatrix}$$

b) We first note that $\mathbf{B}$ has $\boxed{}$ columns, and $\mathbf{A}$ has $\boxed{}$ rows. Therefore, we

can perform this multiplication / cannot perform this multiplication .

$$\mathbf{BA} = \begin{bmatrix} 1 & 6 \\ 3 & -5 \\ -2 & 4 \end{bmatrix} \cdot \begin{bmatrix} 3 & 1 & -1 \\ 2 & 0 & 3 \end{bmatrix}$$

$$= \begin{bmatrix} 1 \cdot 3 + \boxed{} \cdot 2 & 1 \cdot 1 + 6 \cdot \boxed{} & 1 \cdot (-1) + \boxed{} \cdot \boxed{} \\ 3 \cdot \boxed{} + (-5) \cdot \boxed{} & 3 \cdot \boxed{} + (-5) \cdot 0 & \boxed{} \cdot (-1) + (-5) \cdot \boxed{} \\ -2 \cdot 3 + \boxed{} \cdot 2 & \boxed{} \cdot 1 + 4 \cdot \boxed{} & -2 \cdot (-1) + \boxed{} \cdot 3 \end{bmatrix}$$

$$= \begin{bmatrix} 15 & \boxed{} & 17 \\ -1 & 3 & \boxed{} \\ \boxed{} & -2 & \boxed{} \end{bmatrix}$$

c) We first note that **B** has $\boxed{}$ columns, and **C** has $\boxed{}$ rows. Therefore, we

$\underline{\text{can perform this multiplication / cannot perform this multiplication}}$.

$$\mathbf{BC} = \begin{bmatrix} 1 & 6 \\ 3 & -5 \\ -2 & 4 \end{bmatrix} \cdot \begin{bmatrix} 4 & -6 \\ 1 & 2 \end{bmatrix}$$

$$= \begin{bmatrix} 1\cdot4+\boxed{}\cdot1 & \boxed{}\cdot(-6)+6\cdot\boxed{} \\ 3\cdot\boxed{}+(-5)\cdot\boxed{} & 3\cdot(-6)+(-5)\cdot\boxed{} \\ -2\cdot\boxed{}+\boxed{}\cdot1 & \boxed{}\cdot(-6)+\boxed{}\cdot2 \end{bmatrix}$$

$$= \begin{bmatrix} 10 & \boxed{} \\ 7 & -28 \\ \boxed{} & \boxed{} \end{bmatrix}$$

d) We first note that **A** has $\boxed{}$ columns, and **C** has $\boxed{}$ rows. Therefore, we

$\underline{\text{can perform this multiplication / cannot perform this multiplication}}$.

Thus, **AC** $\underline{}$ defined.
$\quad\quad\quad\quad\;\;\underline{\text{is / is not}}$

Example 7 Two of the items sold at Sweet Treats Bakery are gluten-free bagels and gluten-free doughnuts. The table below lists the number of dozens of each product that are sold at the bakery's three stores one week.

	Main Street Store	Avon Road Store	Dalton Avenue Store
Bagels (in dozens)	25	30	20
Doughnuts (in dozens)	40	35	15

The bakery's profit on one dozen bagels is $5, and its profit on one dozen doughnuts is $6. Use matrices to find the total profit on these items at each store for the given week.

We can write the information from the table above in a matrix **S**.

$$\mathbf{S} = \begin{bmatrix} 25 & \boxed{} & 20 \\ 40 & \boxed{} & \boxed{} \end{bmatrix}$$

We can also write the information about the profit in matrix form.

$$\mathbf{P} = \begin{bmatrix} 5 & \boxed{} \end{bmatrix}$$

To find the total profit in each of the three stores, we find the product of the two matrices, **PS**.

$$\mathbf{PS} = \begin{bmatrix} 5 & 6 \end{bmatrix} \cdot \begin{bmatrix} 25 & 30 & \boxed{} \\ \boxed{} & 35 & \boxed{} \end{bmatrix}$$

$$= \begin{bmatrix} 5 \cdot \boxed{} + 6 \cdot \boxed{} & 5 \cdot 30 + \boxed{} \cdot 35 & 5 \cdot \boxed{} + 6 \cdot \boxed{} \end{bmatrix}$$

$$= \begin{bmatrix} 125 + \boxed{} & \boxed{} + 210 & 100 + \boxed{} \end{bmatrix}$$

$$= \begin{bmatrix} 365 & \boxed{} & 190 \end{bmatrix}$$

At the Main Street store the total profit was \$365 for the given week, at the Avon Road store the total profit was $\boxed{}$, and at the Dalton Avenue store the total profit was $\boxed{}$.

Example 8   Write a matrix equation equivalent to the following system of equations:

$$4x + 2y - z = 3,$$
$$9x \quad\quad + z = 5,$$
$$4x + 5y - 2z = 1.$$

First, we can form a matrix of the coefficients from the equations above.

$$\mathbf{A} = \begin{bmatrix} 4 & \boxed{} & -1 \\ 9 & 0 & \boxed{} \\ \boxed{} & 5 & -2 \end{bmatrix}$$

Next we write a variable matrix.

$$\mathbf{X} = \begin{bmatrix} x \\ y \\ z \end{bmatrix}$$

Finally we write a constant matrix.

$$\mathbf{B} = \begin{bmatrix} 3 \\ \boxed{} \\ 1 \end{bmatrix}$$

Matrix equation: $\mathbf{AX} = \boxed{}$

Section 9.5 Inverses of Matrices

Identity Matrix

For any positive integer n, the $n \times n$ **identity matrix** is an $n \times n$ matrix with 1's on the main diagonal and 0's elsewhere and is denoted by

$$\mathbf{I} = \begin{bmatrix} 1 & 0 & 0 & \cdots & 0 \\ 0 & 1 & 0 & \cdots & 0 \\ 0 & 0 & 1 & \cdots & 0 \\ \vdots & \vdots & \vdots & & \vdots \\ 0 & 0 & 0 & \cdots & 1 \end{bmatrix}.$$

Then $\mathbf{AI} = \mathbf{IA} = \mathbf{A}$, for any $n \times n$ matrix $\mathbf{A}$.

Example 1   For

$$\mathbf{A} = \begin{bmatrix} 4 & -7 \\ -3 & 2 \end{bmatrix} \text{ and } \mathbf{I} = \begin{bmatrix} 1 & 0 \\ 0 & 1 \end{bmatrix},$$

find each of the following.

a) **AI**   b) **IA**

a) $\mathbf{AI} = \begin{bmatrix} 4 & -7 \\ -3 & 2 \end{bmatrix} \cdot \begin{bmatrix} 1 & 0 \\ 0 & 1 \end{bmatrix} = \begin{bmatrix} 4 \cdot \square + (-7) \cdot \square & 4 \cdot \square + (-7) \cdot \square \\ \square \cdot 1 + \square \cdot 0 & -3 \cdot \square + \square \cdot 1 \end{bmatrix}$

$= \begin{bmatrix} 4 & \square \\ \square & 2 \end{bmatrix}$

b) $\mathbf{IA} = \begin{bmatrix} 1 & 0 \\ 0 & 1 \end{bmatrix} \cdot \begin{bmatrix} 4 & -7 \\ -3 & 2 \end{bmatrix} = \begin{bmatrix} \square & -7 \\ -3 & \square \end{bmatrix}$

Therefore, $\mathbf{AI} = \mathbf{IA} = \boxed{}$.

Inverse of a Matrix

For an $n \times n$ matrix $\mathbf{A}$, if there is a matrix $\mathbf{A}^{-1}$ for which $\mathbf{A}^{-1} \cdot \mathbf{A} = \mathbf{I} = \mathbf{A} \cdot \mathbf{A}^{-1}$, then $\mathbf{A}^{-1}$ is the **inverse** of $\mathbf{A}$.

Example 2   Verify that

$$\mathbf{B} = \begin{bmatrix} 4 & -3 \\ 3 & -2 \end{bmatrix} \text{ is the inverse of } \mathbf{A} = \begin{bmatrix} -2 & 3 \\ -3 & 4 \end{bmatrix}.$$

Show that $\mathbf{BA} = \mathbf{I} = \mathbf{AB}$.

$$\mathbf{BA} = \begin{bmatrix} 4 & -3 \\ 3 & -2 \end{bmatrix} \cdot \begin{bmatrix} -2 & 3 \\ -3 & 4 \end{bmatrix} = \begin{bmatrix} 4(-2)+(-3)(-3) & 4(\boxed{})+(-3)(4) \\ 3(-2)+(-2)(\boxed{}) & 3(3)+(-2)(\boxed{}) \end{bmatrix}$$

$$= \begin{bmatrix} 1 & \boxed{} \\ \boxed{} & 1 \end{bmatrix} = \mathbf{I}$$

$$\mathbf{AB} = \begin{bmatrix} -2 & 3 \\ -3 & 4 \end{bmatrix} \cdot \begin{bmatrix} 4 & -3 \\ 3 & -2 \end{bmatrix} = \begin{bmatrix} -2(\boxed{})+3(3) & -2(-3)+\boxed{}(-2) \\ -3(4)+4(\boxed{}) & -3(-3)+4(\boxed{}) \end{bmatrix}$$

$$= \begin{bmatrix} \boxed{} & 0 \\ 0 & \boxed{} \end{bmatrix} = \mathbf{I}$$

To find the inverse of a square matrix **A**, we first form the **augmented matrix** consisting of **A** on the left side and the identity matrix with the same dimensions on the right side. The goal is to use row-equivalent operations in order to transform this matrix to an augmented matrix with the identity matrix on the left side. Then, the inverse of **A** is the matrix on the right side. For example,

$$\mathbf{A} = \begin{bmatrix} -2 & 3 \\ -3 & 4 \end{bmatrix} \xrightarrow[\text{matrix}]{\text{augment}} \begin{bmatrix} -2 & 3 & | & 1 & 0 \\ -3 & 4 & | & 0 & 1 \end{bmatrix} \xrightarrow[\text{operations}]{\text{row}} \begin{bmatrix} 1 & 0 & | & a & b \\ 0 & 1 & | & c & d \end{bmatrix} \longrightarrow \mathbf{A}^{-1} = \begin{bmatrix} a & b \\ c & d \end{bmatrix}.$$

Example 3   Find $\mathbf{A}^{-1}$, where

$$\mathbf{A} = \begin{bmatrix} -2 & 3 \\ -3 & 4 \end{bmatrix}.$$

We start by forming the augmented matrix.

$$\begin{bmatrix} -2 & \boxed{} & | & 1 & \boxed{} \\ -3 & \boxed{} & | & \boxed{} & 1 \end{bmatrix}$$

Our goal is to transform this matrix to one of the form

$$\begin{bmatrix} 1 & 0 & | & a & b \\ 0 & 1 & | & c & d \end{bmatrix}.$$

First, we will get a 1 in the upper left corner of the matrix.

$$\left[\begin{array}{cc|cc} 1 & -3/2 & -1/2 & \Box \\ -3 & 4 & \Box & \Box \end{array}\right] \qquad \text{New row } 1 = -\frac{1}{2}(\text{row } 1)$$

Next, we want a 0 in the entry in the lower left corner.

$$\left[\begin{array}{cc|cc} 1 & -3/2 & -1/2 & 0 \\ 0 & -1/2 & \Box & \Box \end{array}\right] \qquad \text{New row } 2 = 3(\text{row } 1) + \text{row } 2$$

Now we want a 1 in row 2, column 2.

$$\left[\begin{array}{cc|cc} 1 & \Box & -1/2 & 0 \\ 0 & 1 & \Box & \Box \end{array}\right] \qquad \text{New row } 2 = -2(\text{row } 2)$$

Finally, we want a 0 in row 1, column 2.

$$\left[\begin{array}{cc|cc} 1 & 0 & \Box & -3 \\ 0 & 1 & 3 & \Box \end{array}\right] \qquad \text{New row } 1 = \frac{3}{2}(\text{row } 2) + \text{row } 1$$

Thus,

$$\mathbf{A}^{-1} = \left[\begin{array}{cc} 4 & \Box \\ \Box & -2 \end{array}\right].$$

The $\boxed{x^{-1}}$ key on a graphing calculator can also be used to find the inverse of a matrix.

Example 4   Find $\mathbf{A}^{-1}$, where

$$\mathbf{A} = \begin{bmatrix} 1 & 2 & -1 \\ 3 & 5 & 3 \\ 2 & 4 & 3 \end{bmatrix}.$$

We start by forming the augmented matrix.

$$\left[\begin{array}{ccc|ccc} 1 & 2 & -1 & 1 & \Box & \Box \\ 3 & 5 & 3 & \Box & 1 & \Box \\ 2 & 4 & 3 & \Box & \Box & 1 \end{array}\right]$$

Our goal is to transform this matrix to one of the form

$$\left[\begin{array}{ccc|ccc} 1 & 0 & 0 & a & b & c \\ 0 & 1 & 0 & d & e & f \\ 0 & 0 & 1 & g & h & i \end{array}\right].$$

$$-3R_1 + R_2 \rightarrow \begin{bmatrix} 1 & 2 & -1 & | & 1 & 0 & 0 \\ \boxed{} & -1 & 6 & | & \boxed{} & 1 & 0 \\ 0 & \boxed{} & 5 & | & -2 & \boxed{} & 1 \end{bmatrix}$$
$$-2R_1 + R_3 \rightarrow$$

$$\frac{1}{5}R_3 \rightarrow \begin{bmatrix} 1 & 2 & -1 & | & 1 & 0 & 0 \\ 0 & -1 & 6 & | & -3 & 1 & 0 \\ 0 & 0 & \boxed{} & | & -2/5 & 0 & \boxed{} \end{bmatrix}$$

$$R_3 + R_1 \rightarrow \begin{bmatrix} 1 & 2 & \boxed{} & | & 3/5 & \boxed{} & 1/5 \\ 0 & -1 & \boxed{} & | & -3/5 & 1 & -6/5 \\ 0 & 0 & 1 & | & -2/5 & 0 & 1/5 \end{bmatrix}$$
$$-6R_3 + R_2 \rightarrow$$

$$2R_2 + R_1 \rightarrow \begin{bmatrix} 1 & 0 & \boxed{} & | & -3/5 & \boxed{} & -11/5 \\ 0 & -1 & 0 & | & -3/5 & 1 & -6/5 \\ 0 & 0 & 1 & | & -2/5 & 0 & 1/5 \end{bmatrix}$$

$$-1 \cdot R_2 \rightarrow \begin{bmatrix} 1 & 0 & 0 & | & -3/5 & 2 & -11/5 \\ 0 & \boxed{} & 0 & | & 3/5 & -1 & 6/5 \\ 0 & 0 & 1 & | & -2/5 & 0 & 1/5 \end{bmatrix}$$

We have

$$\mathbf{A}^{-1} = \begin{bmatrix} -3/5 & 2 & -11/5 \\ 3/5 & -1 & 6/5 \\ -2/5 & 0 & 1/5 \end{bmatrix}.$$

We can write a system of n linear equations in n variables as a matrix equation $\mathbf{AX} = \mathbf{B}$. If $\mathbf{A}$ has an inverse, then the system of equations has a unique solution that can be found by solving for $\mathbf{X}$, as follows:

$$\mathbf{AX} = \mathbf{B}$$
$$\mathbf{A}^{-1}(\mathbf{AX}) = \mathbf{A}^{-1}\mathbf{B}$$
$$(\mathbf{A}^{-1}\mathbf{A})\mathbf{X} = \mathbf{A}^{-1}\mathbf{B}$$
$$\mathbf{IX} = \mathbf{A}^{-1}\mathbf{B}$$
$$\mathbf{X} = \mathbf{A}^{-1}\mathbf{B}.$$

Example 5 Use an inverse matrix to solve the following system of equations:
$$-2x + 3y = 4,$$
$$-3x + 4y = 5.$$

First, we can write this as a matrix equation $\mathbf{AX} = \mathbf{B}$.

$$\begin{bmatrix} -2 & 3 \\ \Box & \Box \end{bmatrix} \cdot \begin{bmatrix} x \\ y \end{bmatrix} = \begin{bmatrix} 4 \\ \Box \end{bmatrix}.$$

Then $\mathbf{X} = \mathbf{A}^{-1}\mathbf{B}$. Earlier, we found $\mathbf{A}^{-1}$. We have

$$\mathbf{A}^{-1} = \begin{bmatrix} 4 & -3 \\ 3 & -2 \end{bmatrix}.$$

Then

$$\begin{bmatrix} x \\ y \end{bmatrix} = \begin{bmatrix} 4 & -3 \\ 3 & -2 \end{bmatrix} \cdot \begin{bmatrix} 4 \\ \Box \end{bmatrix} = \begin{bmatrix} 1 \\ \Box \end{bmatrix}.$$

Thus, the solution of the system is $\left(\Box , \Box \right)$.

Section 9.6   Determinants and Cramer's Rule

With every square matrix, we associate a number called its determinant.

Determinant of a 2×2 Matrix

The **determinant** of the matrix $\begin{bmatrix} a & c \\ b & d \end{bmatrix}$ is denoted $\begin{vmatrix} a & b \\ c & d \end{vmatrix}$ and is defined as

$$\begin{vmatrix} a & b \\ c & d \end{vmatrix} = ad - bc.$$

Example 1   Evaluate $\begin{vmatrix} \sqrt{2} & -3 \\ -4 & -\sqrt{2} \end{vmatrix}$.

$$\begin{vmatrix} \sqrt{2} & -3 \\ -4 & -\sqrt{2} \end{vmatrix} = \left(\sqrt{2}\right)\left(-\sqrt{2}\right) - \left(\boxed{}\right)(-3)$$

$$= \boxed{} - (12) = \boxed{}$$

Often we first find minors and cofactors of matrices in order to evaluate determinants.

Minor

For a square matrix $\mathbf{A} = \left[a_{ij} \right]$, the **minor** M_{ij} of an entry a_{ij} is the determinant of the matrix formed by deleting the ith row and the jth column of $\mathbf{A}$.

Example 2   For the matrix

$$\mathbf{A} = \left[a_{ij} \right] = \begin{bmatrix} -8 & 0 & 6 \\ 4 & -6 & 7 \\ -1 & -3 & 5 \end{bmatrix},$$

find each of the following.

a) M_{11} b) M_{23}

a) The minor M_{11} is the determinant of the matrix formed by eliminating the first row and the first column of $\mathbf{A}$.

$$M_{11} = \begin{vmatrix} -6 & 7 \\ -3 & 5 \end{vmatrix} = -6 \cdot \boxed{} - (-3) \cdot \boxed{}$$

$$= \boxed{} - (-21) = -30 + 21 = \boxed{}$$

b) The minor M_{23} is the determinant of the matrix formed by eliminating the second row and third column of **A**.

$$M_{23} = \begin{vmatrix} -8 & \boxed{} \\ -1 & \boxed{} \end{vmatrix} = \boxed{} \cdot (-3) - (-1) \cdot \boxed{}$$

$$= 24 - \boxed{} = \boxed{}$$

Cofactor

For a square matrix $\mathbf{A} = \begin{bmatrix} a_{ij} \end{bmatrix}$, the **cofactor** A_{ij} of an entry a_{ij} is given by

$$A_{ij} = (-1)^{i+j} M_{ij},$$

where M_{ij} is the minor of a_{ij}.

Example 3   For the matrix

$$\mathbf{A} = \begin{bmatrix} a_{ij} \end{bmatrix} = \begin{bmatrix} -8 & 0 & 6 \\ 4 & -6 & 7 \\ -1 & -3 & 5 \end{bmatrix}$$

find each of the following.

a) A_{11} b) A_{23}

a) $A_{11} = (-1)^{1+1} \cdot \begin{vmatrix} -6 & 7 \\ -3 & 5 \end{vmatrix} = 1 \cdot \left[(-6) \cdot \boxed{} - (-3) \cdot \boxed{} \right]$

$$= \boxed{} + 21 = \boxed{}$$

b) $A_{23} = (-1)^{2+3} \cdot M_{23} = (-1)^5 \begin{vmatrix} -8 & \boxed{} \\ -1 & \boxed{} \end{vmatrix} = -1 \cdot \left[\boxed{} \cdot (-3) - (-1) \cdot \boxed{} \right]$

$$= -1 \cdot (24) = \boxed{}$$

Determinant of Any Square Matrix

For any square matrix **A** of order $n \times n$ $(n > 1)$, we define the **determinant** of **A**, denoted $|\mathbf{A}|$, as follows. Choose any row or column. Multiply each element in that row or column by its cofactor and add the results. The determinant of a 1×1 matrix is simply the element of the matrix. The value of a determinant will be the same no matter which row or column is chosen.

Example 4 Evaluate $|\mathbf{A}|$ by expanding across the third row.

$$\mathbf{A} = \begin{bmatrix} -8 & 0 & 6 \\ 4 & -6 & 7 \\ -1 & -3 & 5 \end{bmatrix}$$

The entries in the third row are -1, -3, and 5, so

$$|\mathbf{A}| = (-1)A_{31} + (-3)A_{32} + 5A_{33}$$

$$= (-1)(-1)^{3+1}\begin{vmatrix} 0 & 6 \\ -6 & 7 \end{vmatrix} + (-3)(-1)^{3+2}\begin{vmatrix} -8 & \square \\ 4 & \square \end{vmatrix} + 5(-1)^{3+3}\begin{vmatrix} -8 & \square \\ 4 & \square \end{vmatrix}$$

$$= -1 \cdot \left[0 \cdot 7 - (-6) \cdot \square \right] + 3 \cdot \left[-8 \cdot \square - 4 \cdot \square \right] + 5 \cdot \left[\square \cdot (-6) - 4 \cdot \square \right]$$

$$= -36 - \square + 240 = \square$$

Example 5 Use a graphing calculator to evaluate $|\mathbf{A}|$:

$$\mathbf{A} = \begin{bmatrix} 1 & 6 & -1 \\ -3 & -5 & 3 \\ 0 & 4 & 2 \end{bmatrix}.$$

Enter **A**. Then select the operation det from the MATRIX MATH menu, enter the name of the matrix, **A**, and press ENTER. We see that $|\mathbf{A}| = \square$.

Cramer's Rule for 2×2 Systems

The solution of the system of equations

$$a_1 x + b_1 y = c_1,$$
$$a_2 x + b_2 y = c_2$$

is given by

$$x = \frac{D_x}{D}, \quad y = \frac{D_y}{D},$$

where

$$D = \begin{vmatrix} a_1 & b_1 \\ a_2 & b_2 \end{vmatrix}, \quad D_x = \begin{vmatrix} c_1 & b_1 \\ c_2 & b_2 \end{vmatrix}, \quad D_y = \begin{vmatrix} a_1 & c_1 \\ a_2 & c_2 \end{vmatrix}, \text{ and } D \neq 0.$$

Example 6 Solve using Cramer's rule:
$$2x + 5y = 7,$$
$$5x - 2y = -3.$$

First we need to calculate D, D_x, and D_y.

$$D = \begin{vmatrix} 2 & \boxed{} \\ 5 & \boxed{} \end{vmatrix} = -4 - 25 = \boxed{}$$

$$D_x = \begin{vmatrix} 7 & 5 \\ \boxed{} & -2 \end{vmatrix} = -14 + \boxed{} = 1$$

$$D_y = \begin{vmatrix} 2 & \boxed{} \\ 5 & \boxed{} \end{vmatrix} = -6 - 35 = \boxed{}$$

Cramer's rule tells us that

$$x = \frac{D_x}{D} = \frac{1}{-29} = -\frac{\boxed{}}{\boxed{}}, \quad y = \frac{D_y}{D} = \frac{-41}{-29} = \frac{\boxed{}}{\boxed{}}.$$

The solution of the system of equations is $\left(-\dfrac{1}{29}, \boxed{} \right)$.

We can also use a graphing calculator to solve this system of equations using Cramer's rule.

Cramer's Rule for 3×3 Systems

The solution of the system of equations

$$a_1x + b_1y + c_1z = d_1,$$
$$a_2x + b_2y + c_2z = d_2,$$
$$a_3x + b_3y + c_3z = d_3,$$

is given by

$$x = \frac{D_x}{D}, \quad y = \frac{D_y}{D}, \quad z = \frac{D_z}{D},$$

where

$$D = \begin{vmatrix} a_1 & b_1 & c_1 \\ a_2 & b_2 & c_2 \\ a_3 & b_3 & c_3 \end{vmatrix}, \quad D_x = \begin{vmatrix} d_1 & b_1 & c_1 \\ d_2 & b_2 & c_2 \\ d_3 & b_3 & c_3 \end{vmatrix}, \quad D_y = \begin{vmatrix} a_1 & d_1 & c_1 \\ a_2 & d_2 & c_2 \\ a_3 & d_3 & c_3 \end{vmatrix}, \quad D_z = \begin{vmatrix} a_1 & b_1 & d_1 \\ a_2 & b_2 & d_2 \\ a_3 & b_3 & d_3 \end{vmatrix}, \text{ and } D \neq 0.$$

Example 7 Solve using Cramer's rule:

$$x - 3y + 7z = 13,$$
$$x + \ y + \ z = 1,$$
$$x - 2y + 3z = 4.$$

$$D = \begin{vmatrix} 1 & \Box & 7 \\ \Box & 1 & 1 \\ 1 & -2 & -3 \end{vmatrix} = -10 \qquad\qquad D_y = \begin{vmatrix} 1 & \Box & 7 \\ 1 & 1 & 1 \\ 1 & \Box & 3 \end{vmatrix} = -6$$

$$D_x = \begin{vmatrix} 13 & -3 & \Box \\ 1 & \Box & 1 \\ \Box & -2 & 3 \end{vmatrix} = 20 \qquad\qquad D_z = \begin{vmatrix} 1 & -3 & 13 \\ 1 & 1 & 1 \\ \Box & -2 & \Box \end{vmatrix} = -24$$

$$x = \frac{D_x}{D} = \frac{20}{\Box} = \Box, \qquad y = \frac{D_y}{D} = \frac{\Box}{-10} = \frac{3}{5}, \qquad z = \frac{D_z}{D} = \frac{\Box}{-10} = \frac{12}{\Box}$$

Section 9.7 Systems of Inequalities and Linear Programming

> A **linear inequality in two variables** is an inequality that can be written in the form
> $$Ax + By < C,$$
> where A, B, and C, are real numbers and A and B are not both zero. The symbol $<$ can be replaced with $\leq$, $>$, or $\geq$.

The **graph of an inequality** represents its **solution set**; that is, the set of all ordered pairs that make it true. When graphing linear inequalities, we first graph the **related equation** which is a line that divides the coordinate plane into two half-planes. One of these half-planes satisfies the inequality. Either *all* points in a half-plane are in the solution set or *none* are.

> To graph a linear inequality in two variables:
> 1. Replace the inequality symbol with an equals sign and graph this related equation. If the inequality symbol is $<$ or $>$, draw the line dashed. If the inequality symbol is $\leq$ or $\geq$, draw the line solid.
> 2. The graph consists of a half-plane on one side of the line and, if the line is solid, the line as well. To determine which half-plane to shade, test a point not on the line in the original inequality. If that point is a solution, shade the half-plane containing that point. If not, shade the opposite half-plane.

Example 1   Graph: $y < x + 3$.

The related equation is $y = x + 3$. The intercepts are $(-3, 0)$ and $(0, 3)$. We use a

_____ line because the inequality symbol is $<$. Next we test a point not on the line
solid / dashed

to determine which half-plane to shade. We will use $(0, 0)$.

$$\frac{y < x + 3}{0 \,?\, 0 + 3}$$

$$0 \;\;\boxed{\phantom{<}}\quad \text{TRUE}$$

This tells us that the half-plane that contains $(0, 0)$ is in the solution set of the inequality, so we shade it.

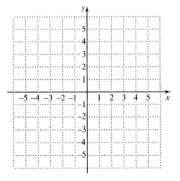

Examples 2   Graph: $3x + 4y \geq 12$.

The related equation is $3x + 4y = 12$. The intercepts are $(0, 3)$ and $(4, 0)$. We use a
_____ line because the inequality symbol is $\geq$. Next we use a test point not on
solid / dashed
the line to determine which half-plane to shade. We will use $(0, 0)$.

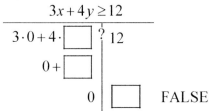

$(0, 0)$ is not a solution, so all the points in the half-plane that does *not* contain $(0, 0)$ are
solutions. We shade that region.

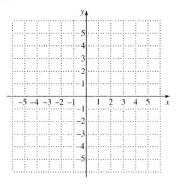

Example 3   Graph $x > -3$ on a plane.

The related equation is $x = \boxed{}$. We graph it using a _____ line because the
solid / dashed
inequality symbol is $>$. We test $(0, 0)$ to determine which half-plane to shade. Since $0 > -3$
is _____, we shade the half-plane containing $(0, 0)$.
true / false

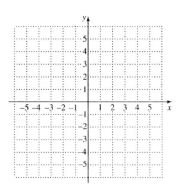

Example 4   Graph $y \leq 4$ on a plane.

First graph the related equation $y = 4$ using a _____ line. We shade the half-plane
 solid / dashed

that contains ordered pairs for which $y \leq 4$. This is the half-plane _____ the line.
 above / below

We can also test a point not on the line. We use $(-2, 5)$.

$$\frac{y \leq 4}{\boxed{} \; ? \; 4} \quad \text{FALSE}$$

Because $(-2, 5)$ is not a solution, we shade the half-plane that does not contain that point.

A system of inequalities consists of two or more inequalities considered simultaneously.
To graph a system of linear inequalities, we graph each inequality and determine the
region that is common to *all* the solution sets.

Example 5   Graph the solution set of the system

$$x + y \leq 4,$$
$$x - y \geq 2.$$

We graph $x + y \leq 4$ by first graphing the related equation $x + y = \boxed{}$. The intercepts are

$(0, 4)$ and $\left(\boxed{}, 0\right)$. We use a solid line. When we test $(0, 0)$ we find that it _____ a
 is / is not

solution of the inequality, so we shade the half-plane containing $(0, 0)$.

Next we graph $x - y = 2$. The intercepts are $(0, -2)$ and $\left(2, \boxed{}\right)$. We use a solid line.

We find that $(0, 0)$ _____ a solution, so we shade the half-plane that does not contain
 is / is not

$(0, 0)$. The region where the two solution sets overlap is the solution set of the system of

inequalities.

> The solution set of a system of inequalities may consist of a polygon and its interior.

Example 6   Graph the following system in inequalities and find the coordinates of any vertices formed:

$$3x - y \leq 6, \quad (1)$$
$$y - 3 \leq 0, \quad (2)$$
$$x + y \geq 0. \quad (3)$$

For inequality (1): Graph the related equation $3x - y = 6$ using a _____ line. The
                                                                                              solid / dashed

intercepts are $\left(0, \boxed{}\right)$ and $\left(2, \boxed{}\right)$. We find that $(0, 0)$ _____ a solution, so we
                                                                                              is / is not

will shade to the left of the line.

For inequality (2): Graph the related equation $y - 3 = 0$, or $y = 3$, using a

_____ line. This is a horizontal line that passes through $(0, 3)$. We find that
 solid / dashed

$(0, 0)$ _____ a solution, so we will shade below the line.
          is / is not

For inequality (3): Graph the related equation $x + y = 0$. Two points on the line are $(0, 0)$

and $\left(3, \boxed{}\right)$. We test $(1, 1)$ and find that it is a solution, so we will shade above the line.

The solution set of the system of inequalities is the intersection of the three individual solution sets. We shade this polygon and its interior.

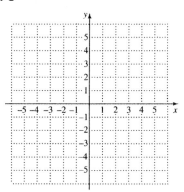

Now we find the coordinates of the vertices by solving three systems of related equations.

For the system from inequalities (1) and (2), we have $\left(3, \boxed{}\right)$.

For the system from inequalities (1) and (3), we have $\left(\dfrac{3}{2}, \boxed{}\right)$.

For the system from inequalities (2) and (3), we have $\left(\boxed{}, 3\right)$.

Linear programming allows us to find the maximum or minimum value of a linear **objective function** subject to several **constraints** which are expressed as inequalities. The solution set of the system of inequalities composed of the constraints contains all the **feasible solutions**.

It can be shown that the maximum and minimum values of the objective function occur at a vertex of the region of feasible solutions.

Linear Programming Procedure

To find the maximum or minimum of a linear objective function, subject to constraints:

1. Graph the region of feasible solutions.
2. Determine the coordinates of the vertices of the region.
3. Evaluate the objective function at each vertex. The largest and smallest of those values are the maximum and minimum values of the objective function, respectively.

Example 7 *Maximizing Profit.* Aspen Carpentry makes bookcases and desks. Each bookcase requires 5 hr of woodworking and 4 hr of finishing. Each desk requires 10 hr of woodworking and 3 hr of finishing. Each month the shop has 600 hr of labor available for woodworking and 240 hr for finishing. The profit on each bookcase is $75 and on each desk is $140. How many of each product should be made each month in order to maximize profit? What is the maximum profit?

Let $x =$ the number of bookcases to be produced and sold and let $y =$ the number of desks to be produced and sold.

The profit is given by

$$P = \boxed{} \cdot x + \boxed{} \cdot y.$$

Next we consider the constraints. The total number of hours of woodworking on bookcases and desks must satisfy

$$5x + 10y \leq \boxed{}.$$

The total number of hours of finishing on bookcases and desks must satisfy

$$4x + \boxed{} y \leq 240.$$

Finally we have $x \geq 0$ and $y \geq 0$.

Maximize

$$P = 75x + \boxed{} y$$

subject to

$$5x + \boxed{} y \leq 600,$$
$$4x + 3y \leq \boxed{},$$
$$x \geq 0,$$
$$y \geq 0.$$

We graph the system of inequalities and find the vertices.

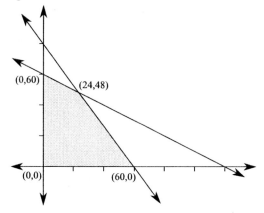

The maximum value will occur at a vertex, so we check the value of P at each one.

Vertex (x, y)	Profit $P = \boxed{} x + \boxed{} y$		
$(0, 0)$	$75 \cdot \boxed{} + 140 \cdot \boxed{} = \boxed{}$		
$(60, 0)$	$75 \cdot \boxed{} + 140 \cdot 0 = \boxed{}$		
$(24, 48)$	$75 \cdot 24 + 140 \cdot \boxed{} = 8520$		
$(0, 60)$	$75 \cdot \boxed{} + 140 \cdot \boxed{} = 8400$		

We see that the maximum profit of $\boxed{}$ occurs when 24 bookcases and $\boxed{}$ decks are produced and sold.

Section 9.8   Partial Fractions

When we write a **partial fraction decomposition** of a rational expression, we write the rational expression as a sum of two or more simpler rational expressions.

Procedure for Decomposing a Rational Expression into Partial Fractions

Consider a rational expression $P(x)/Q(x)$ such that $P(x)$ and $Q(x)$ have no common factor other than 1 or -1.

1. If the degree of $P(x)$ is greater than or equal to the degree of $Q(x)$, divide to express $P(x)/Q(x)$ as a quotient $+$ remainder $/Q(x)$ and follow steps $(2)-(5)$ to decompose the resulting rational expression.

2. If the degree of $P(x)$ is less than the degree of $Q(x)$, factor $Q(x)$ into linear factors of the form $(px+q)^n$ and/or quadratic factors of the form $(ax^2+bx+c)^m$. Any quadratic factor ax^2+bx+c must be irreducible, meaning it cannot be factored into linear factors with rational coefficients.

3. Assign to each linear factor $(px+q)^n$ the sum of n partial fractions:

$$\frac{A_1}{px+q} + \frac{A_2}{(px+q)^2} + \cdots + \frac{A_n}{(px+q)^n}.$$

4. Assign to each quadratic factor $(ax^2+bx+c)^m$ the sum of m partial fractions:

$$\frac{B_1 x + C_1}{ax^2+bx+c} + \frac{B_2 x + C_2}{(ax^2+bx+c)^2} + \cdots + \frac{B_m x + C_m}{(ax^2+bx+c)^m}.$$

5. Apply algebraic methods to find the constants in the numerators of the partial fractions.

Example 1   Decompose into partial fractions:

$$\frac{4x-13}{2x^2+x-6}.$$

The degree of the numerator, 1, is less than the degree of the denominator, 2, so we factor the denominator.

$$2x^2+x-6=(x+2)(\boxed{})$$

We can decompose the fraction as

$$\frac{4x-13}{(x+2)(2x-3)} = \frac{A}{x+2} + \frac{B}{2x-3}.$$

We add the two expressions on the right.

$$\frac{4x-13}{(x+2)(2x-3)} = \frac{A(2x-3)+B(\boxed{})}{(x+2)(2x-3)}$$

Next we equate the numerators.

$$4x-13 = A(2x-3)+B(x+2)$$

Since this is true for all x, we can choose any value of x and still have a true equation.

If we let $x = \dfrac{3}{2}$, then $2x-3=0$ and A will be eliminated.

$$4 \cdot \frac{3}{2} - 13 = A\left(2 \cdot \boxed{} - 3\right) + B\left(\frac{3}{2}+2\right)$$

$$-7 = 0 + \boxed{}\, B$$

$$-2 = B \qquad \text{Multiplying by } \frac{2}{7}$$

Now let's choose an x-value that will allow us to eliminate B. When $x=-2$, $x+2=0$.

$$4(-2)-13 = A(2(-2)-3)+B(\boxed{}+2)$$

$$-21 = -7A + \boxed{}$$

$$3 = A \qquad \text{Dividing by } -7$$

We have

$$\frac{4x-13}{2x^2+x-6} = \frac{\boxed{}}{x+2} + \frac{\boxed{}}{2x-3}, \text{ or } \frac{3}{x+2} - \frac{\boxed{}}{2x-3}.$$

Example 2   Decompose into partial fractions:

$$\frac{7x^2 - 29x + 24}{(2x-1)(x-2)^2}.$$

The degree of the numerator, 2, is less than the degree of the denominator, 3. We can decompose the fraction as

$$\frac{7x^2 - 29x + 24}{(2x-1)(x-2)^2} = \frac{A}{2x-1} + \frac{B}{x-2} + \frac{C}{(x-2)^2}.$$

We add the three expressions on the right.

$$\frac{7x^2 - 29x + 24}{(2x-1)(x-2)^2} = \frac{A(x-2)^2 + B(2x-1)(x-2) + C\left(\boxed{}\right)}{(2x-1)(x-2)^2}$$

Now we equate the numerators.

$$7x^2 - 29x + 24 = A(x-2)^2 + B(2x-1)(x-2) + C(2x-1)$$

If we let $x = \frac{1}{2}$, we can eliminate B and C.

$$7\left(\frac{1}{2}\right)^2 - 29 \cdot \frac{1}{2} + \boxed{} = A\left(\frac{1}{2} - 2\right)^2 + 0 + 0$$

$$\frac{7}{4} - \frac{29}{2} + 24 = \frac{9}{4} A$$

$$\frac{7}{4} - \frac{\boxed{}}{4} + \frac{\boxed{}}{4} = \frac{9}{4} A$$

$$\frac{45}{4} = \frac{9}{4} A$$

$$\boxed{} = A \qquad \text{Multiplying by } \frac{4}{9}$$

When $x = 2$, we can eliminate A and B.

$$7 \cdot 2^2 - 29 \cdot \boxed{} + 24 = 0 + 0 + C(2 \cdot 2 - 1)$$

$$28 - \boxed{} + 24 = 3C$$

$$-6 = 3C$$

$$\boxed{} = C \qquad \text{Dividing by 3}$$

Now we substitute 5 for A and -2 for C and choose a value of x that will allow us to find B. We will use $x = 1$.

$$7 \cdot 1^2 - 29 \cdot \boxed{} + 24 = 5(1-2)^2 + B\left(2 \cdot \boxed{} - 1\right)(1-2) + (-2)(2 \cdot 1 - 1)$$

$$7 - \boxed{} + 24 = 5 \cdot 1 + B(-1) + (-2)$$

$$2 = \boxed{} - B - \boxed{}$$

$$2 = \boxed{} - B$$

$$\boxed{} = -B$$

$$1 = B$$

We have

$$\frac{7x^2 - 29x + 24}{(2x-1)(x-2)^2} = \frac{5}{2x-1} + \frac{\boxed{}}{x-2} - \frac{2}{(x-2)^2}.$$

Example 3   Decompose into partial fractions:

$$\frac{6x^3 + 5x^2 - 7}{3x^2 - 2x - 1}.$$

The degree of the numerator, 3, is greater than the degree of the denominator, 2, so we divide to find an equivalent expression.

$$
\begin{array}{r}
2x + 3 \\
3x^2 - 2x - 1\overline{)6x^3 + 5x^2 - 7} \\
\underline{6x^3 - \boxed{} - 2x} \\
9x^2 + \boxed{} - 7 \\
\underline{9x^2 - 6x - \boxed{}} \\
8x - 4
\end{array}
$$

Then $\dfrac{6x^3 + 5x^2 - 7}{3x^2 - 2x - 1} = 2x + \boxed{} + \dfrac{8x - 4}{3x^2 - 2x - 1}.$

Now we find the partial fraction decomposition of

$$\frac{8x - 4}{3x^2 - 2x - 1}, \text{ or } \frac{8x - 4}{(3x+1)(x-1)}.$$

$$\frac{8x - 4}{(3x+1)(x-1)} = \frac{A}{3x+1} + \frac{B}{\boxed{}}$$

Multiply both sides by $(3x+1)(x-1)$.

$$8x - 4 = \boxed{}(x-1) + \boxed{}(3x+1)$$

Let $x = 1$ to find B:

$$8 - 4 = A \cdot 0 + B(4)$$

$$\boxed{} = 4B$$

$$1 = B, \text{ or } B = 1$$

Now let $x = -\dfrac{1}{3}$ to find A:

$$-\dfrac{8}{3} - \dfrac{12}{3} = A\left(-\dfrac{1}{3} - \boxed{}\right) + B \cdot 0$$

$$-\dfrac{20}{3} = -\dfrac{4}{3}A$$

$$-20 = -4A$$

$$\boxed{} = A$$

Then $\dfrac{6x^3 + 5x^2 - 7}{3x^2 - 2x - 1} = 2x + 3 + \dfrac{\boxed{}}{3x + 1} + \dfrac{\boxed{}}{x - 1}.$

Example 4   Decompose into partial fractions:

$$\dfrac{7x^2 - 29x + 24}{(2x - 1)(x - 2)^2}.$$

The degree of the numerator, 2, is less than the degree of the denominator, 3, so we decompose the fraction as

$$\dfrac{7x^2 - 29x + 24}{(2x - 1)(x - 2)^2} = \dfrac{A}{2x - 1} + \dfrac{B}{x - 2} + \dfrac{C}{(x - 2)^2}.$$

We add the three expressions on the right.

$$\dfrac{7x^2 - 29x + 24}{(2x - 1)(x - 2)^2} = \dfrac{A(x - 2)^2 + \boxed{}(2x - 1)(x - 2) + C\left(2x - \boxed{}\right)}{(2x - 1)(x - 2)^2}$$

Now we equate the numerators.

$$7x^2 - 29x + \boxed{} = \boxed{}(x - 2)^2 + B(2x - 1)(x - 2) + C(2x - 1)$$

$$= A\left(x^2 - 4x + \boxed{}\right) + B\left(2x^2 - 5x + \boxed{}\right) + C(2x - 1)$$

$$= Ax^2 - 4Ax + \boxed{}A + 2Bx^2 - 5Bx + \boxed{}B + 2Cx - \boxed{}$$

$$= (A + 2B)x^2 + \left(-4A - \boxed{}B + 2C\right)x + (4A + 2B - C)$$

Next we equate corresponding coefficients.

$$7 = A + 2B$$
$$\boxed{} = -4A - 5B + 2C$$
$$24 = 4A + 2B - \boxed{}$$

We have a system of three equations. You should confirm that the solution is
$$A = 5, \ B = 1, \text{ and } C = -2.$$

Then we have

$$\frac{7x^2 - 29x + 24}{(2x - 1)(x - 2)^2} = \frac{\boxed{}}{2x - 1} + \frac{\boxed{}}{x - 2} - \frac{2}{(x - 2)^2}.$$

Example 5   Decompose into partial fractions:

$$\frac{11x^2 - 8x - 7}{(2x^2 - 1)(x - 3)}.$$

$$\frac{11x^2 - 8x - 7}{(2x^2 - 1)(x - 3)} = \frac{Ax + B}{2x^2 - 1} + \frac{C}{\boxed{}}$$

Multiply both sides by $(2x^2 - 1)(x - 3)$.

$$11x^2 - 8x - 7 = (Ax + B)\left(x - \boxed{}\right) + C(2x^2 - 1)$$
$$= Ax^2 - 3Ax + \boxed{}x - 3B + 2Cx^2 - \boxed{}$$
$$= (A + 2C)x^2 + \left(-3A + \boxed{}\right)x + (-3B - C)$$

Equate coefficients.

$$11 = A + 2C,$$
$$\boxed{} = -3A + B,$$
$$-7 = -3B - \boxed{}$$

Solving this system of equations, we have

$$A = 3, \ B = 1, \text{ and } C = 4.$$

$$\frac{11x^2 - 8x - 7}{(2x^2 - 1)(x - 3)} = \frac{3x + \boxed{}}{2x^2 - 1} + \frac{\boxed{}}{x - 3}$$

Section 10.1 The Parabola

Parabola

A **parabola** is the set of all points in a plane equidistant from a fixed line (the **directrix**) and a fixed point not on the line (the **focus**).

The graph of the quadratic function $f(x) = ax^2 + bx + c$, $a \neq 0$, is a parabola. The line that is perpendicular to the directrix and contains the focus is the **axis of symmetry**. The **vertex** is the midpoint of the segment between the focus and the directrix.

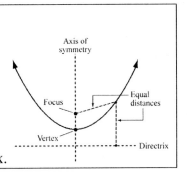

Standard Equation of a Parabola With Vertex $(0,0)$ and Vertical Axis of Symmetry

The standard equation of a parabola with vertex $(0,0)$ and directrix $y = -p$ is

$$x^2 = 4py.$$

The focus is $(0, p)$, and the y-axis is the axis of symmetry. When $p > 0$, the parabola opens up; when $p < 0$, the parabola opens down.

Standard Equation of a Parabola With Vertex $(0,0)$

and Horizontal Axis of Symmetry

The standard equation of a parabola with vertex $(0,0)$ and directrix $x = -p$ is

$$y^2 = 4px.$$

The focus is $(p, 0)$, and the x-axis is the axis of symmetry. When $p > 0$, the parabola opens right; when $p < 0$, the parabola opens to the left.

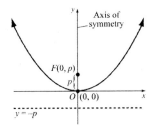

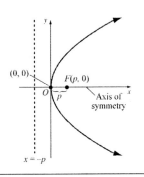

Example 1   Find the vertex, the focus, and the directrix of the parabola $y = -\dfrac{1}{12}x^2$. Then graph the parabola.

$$y = -\frac{1}{12}x^2$$

We want to write this in the form $x^2 = 4py$.

$$x^2 = \boxed{} \cdot y$$

$$x^2 = 4\left(\boxed{}\right)y$$

Vertex: $\left(0, \boxed{}\right)$

Focus $(0, p)$: $\left(0, \boxed{}\right)$

Directrix $y = -p$: $y = \boxed{}$

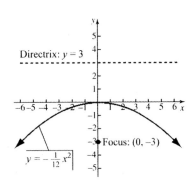

Example 2   Find an equation of the parabola with vertex $(0,0)$ and focus $(5,0)$. Then graph the parabola.

$$y^2 = 4px$$

Focus: $(5,0)$ $p = \boxed{}$

$$y^2 = 4\left(\boxed{}\right)x$$

$$y^2 = \boxed{} \cdot x$$

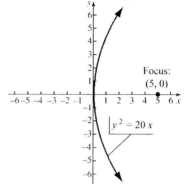

Focus:
$(5,0)$

$y^2 = 20x$

On some graphing calculators, the conics application from the APPS menu can be used to graph parabolas.

Standard Equation of a Parabola with Vertex (h,k) and Vertical Axis of Symmetry

The standard equation of a parabola with vertex (h,k) and vertical axis of symmetry is

$$(x-h)^2 = 4p(y-k),$$

where the vertex is (h,k), the focus is $(h,k+p)$, and the directrix is $y = k-p$.

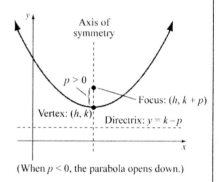

(When $p < 0$, the parabola opens down.)

Standard Equation of a Parabola with Vertex (h,k) and Horizontal Axis of Symmetry

The standard equation of a parabola with vertex (h,k) and horizontal axis of symmetry is

$$(y-k)^2 = 4p(x-h),$$

where the vertex is (h,k), the focus is $(h+p,k)$, and the directrix is $x = h-p$.

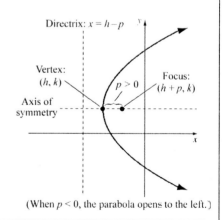

(When $p < 0$, the parabola opens to the left.)

Example 3   For the parabola

$$x^2 + 6x + 4y + 5 = 0,$$

find the vertex, the focus, and the directrix.   Then draw the graph.

$$(x - h)^2 = 4p(y - k)$$

$$x^2 + 6x \qquad = -4y - 5$$

$$x^2 + 6x + \boxed{} = -4y - 5 + \boxed{}$$

$$\left(\boxed{}\right)^2 = -4y + \boxed{}$$

$$(x + 3)^2 = -4(y - 1)$$

$$\left[x - \left(\boxed{}\right)\right]^2 = 4\left(\boxed{}\right)(y - 1)$$

$$h = -3, \qquad k = \boxed{}, \qquad p = \boxed{}$$

Vertex (h, k): $(-3, 1)$

Focus $(h, k + p)$: $\left(-3, \boxed{}\right)$

Directrix $y = k - p$:

$$y = 1 - (-1)$$

$$y = \boxed{}$$

Some points on the graph:

$$\left(-4, \frac{3}{4}\right)$$

$$\left(-2, \frac{3}{4}\right)$$

$$(-5, 0)$$

$$(-1, 0)$$

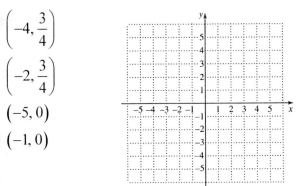

Example 4   For the parabola

$$y^2 - 2y - 8x - 31 = 0,$$

find the vertex, the focus, and the directrix. Then draw the graph.

$$(y-k)^2 = 4p(x-h)$$

$$y^2 - 2y \qquad = 8x + 31$$

$$y^2 - 2y + \boxed{} = 8x + 31 + \boxed{}$$

$$\left(\boxed{}\right)^2 = 8x + \boxed{}$$

$$(y-1)^2 = 8(x+4)$$

$$(y-1)^2 = 4(2)\left[x - \left(\boxed{}\right)\right]$$

$$h = \boxed{}, \qquad k = 1, \qquad p = \boxed{}$$

Vertex (h, k): $(-4, 1)$

Focus $(h + p, k)$: $(-2, 1)$

Directrix $x = h - p$:

$$x = -4 - 2$$

$$x = \boxed{}$$

Some points on the graph:

$$\left(-\frac{31}{8}, 2\right)$$

$$\left(-\frac{31}{8}, 0\right)$$

$$\left(-\frac{7}{2}, 3\right)$$

$$\left(-\frac{7}{2}, -1\right)$$

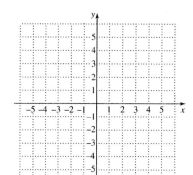

Section 10.2   The Circle and the Ellipse

Standard Equation of a Circle

A **circle** is the set of all points in a plane that are at a fixed distance from a fixed point (the **center**) in the plane.

The standard equation of a circle with center (h, k) and radius r is

$$(x-h)^2 + (y-k)^2 = r^2.$$

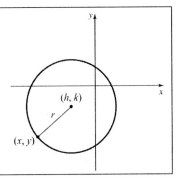

Example 1   For the circle

$$x^2 + y^2 - 16x + 14y + 32 = 0,$$

find the center and the radius. Then graph the circle.

$$(x-h)^2 + (y-k)^2 = r^2$$

$$x^2 + y^2 - 16y + 14y + 32 = 0$$

$$x^2 - 16x \qquad + y^2 + 14y \qquad = -32$$

$$x^2 - 16x + \boxed{} + y^2 + 14y + \boxed{} = -32 + \boxed{} + \boxed{}$$

$$(x-8)^2 + \left(\boxed{}\right)^2 = \boxed{}$$

$$(x-8)^2 + \left[y - (-7)\right]^2 = 9^2$$

Center: $\left(8, \boxed{}\right)$

Radius: $\boxed{}$

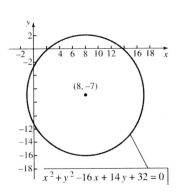

We can also graph circles using the Conics CIRCLE APP on a graphing calculator.

Standard Equation of an Ellipse with Center at the Origin

An **ellipse** is the set of all points in a plane, the sum of whose distances from two fixed points (the **foci**) is constant. The **center** of an ellipse is the midpoint of the segment between the foci.

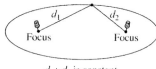

$d_1 + d_2$ is constant.

Major Axis Horizontal

$$\frac{x^2}{a^2} + \frac{y^2}{b^2} = 1, \quad a > b > 0$$

Vertices: $(-a, 0)$, $(a, 0)$

y-intercepts: $(0, -b)$, $(0, b)$

Foci: $(-c, 0)$, $(c, 0)$, where $c^2 = a^2 - b^2$

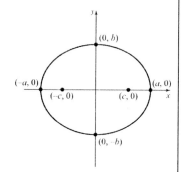

Major Axis Vertical

$$\frac{x^2}{b^2} + \frac{y^2}{a^2} = 1, \quad a > b > 0$$

Vertices: $(0, -a)$, $(0, a)$

x-intercepts: $(-b, 0)$, $(b, 0)$

Foci: $(0, -c)$, $(0, c)$, where $c^2 = a^2 - b^2$

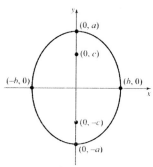

Example 2   Find the standard equation of the ellipse with vertices $(-5, 0)$ and $(5, 0)$ and foci $(-3, 0)$ and $(3, 0)$. Then graph the ellipse.

$$\frac{x^2}{a^2} + \frac{y^2}{b^2} = 1$$

Vertices: $(-5, 0)$, $(5, 0)$ $a = \boxed{}$

Foci: $(-3, 0)$, $(3, 0)$ $c = \boxed{}$

$\quad c^2 = a^2 - b^2$

$\quad 3^2 = \boxed{}^2 - b^2$

$\quad 9 = 25 - b^2$

$\quad b^2 = 16$

$\quad b = \boxed{}$

$$\frac{x^2}{\Box^2} + \frac{y^2}{\Box^2} = 1$$

$$\frac{x^2}{25} + \frac{y^2}{16} = 1$$

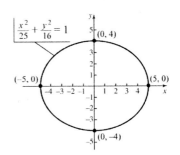

y-intercepts: $\left(0, \boxed{}\right)$, $(0, 4)$

When the equation of an ellipse is written in standard form, we can graph it using the Conics ELLIPSE APP on a graphing calculator.

Example 3 For the ellipse

$$9x^2 + 4y^2 = 36,$$

find the vertices and the foci. Then draw the graph.

$$\frac{9x^2}{\Box} + \frac{4y^2}{36} = \frac{36}{36}$$

$$\frac{x^2}{4} + \frac{y^2}{\Box} = 1$$

$$\frac{x^2}{\Box^2} + \frac{y^2}{3^2} = 1$$

$$a = \boxed{}, \qquad b = 2$$

Vertices: $(0, a)$, $(0, -a)$

$\qquad (0, 3)$, $\left(0, \boxed{}\right)$

Foci: $(0, c)$, $(0, -c)$

$\qquad \left(0, \boxed{}\right)$, $\left(0, -\sqrt{5}\right)$

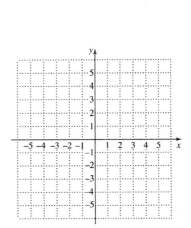

$$c^2 = a^2 - b^2$$

$$c^2 = \boxed{} - 4$$

$$c^2 = 5$$

$$c = \boxed{}$$

x-intercepts: $(-b, 0)$, $(b, 0)$

$\qquad \left(\boxed{}, 0\right)$, $(2, 0)$

Standard Equation of an Ellipse with Center at (h, k)

Major Axis Horizontal

$$\frac{(x-h)^2}{a^2}+\frac{(y-k)^2}{b^2}=1, \quad a>b>0$$

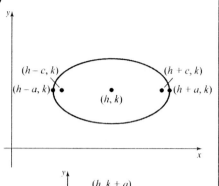

Vertices: $(h-a, k)$, $(h+a, k)$

Length of minor axis: $2b$

Foci: $(h-c, k)$, $(h+c, k)$, where $c^2 = a^2 - b^2$

Major Axis Vertical

$$\frac{(x-h)^2}{b^2}+\frac{(y-k)^2}{a^2}=1, \quad a>b>0$$

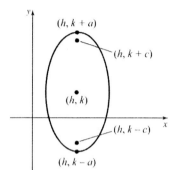

Vertices: $(h, k-a)$, $(h, k+a)$

Length of minor axis: $2b$

Foci: $(h, k-c)$, $(h, k+c)$, where $c^2 = a^2 - b^2$

Example 4   For the ellipse

$$4x^2 + y^2 + 24x - 2y + 21 = 0,$$

find the center, the vertices, and the foci. Then draw the graph.

$$4\left(x^2+6x \quad \right)+\left(y^2-2y \quad \right)=-21$$

$$4\left(x^2+6x+\boxed{}\right)+\left(y^2-2y+1\right)=-21+\boxed{}+1$$

$$4\left(\boxed{}\right)^2+\left(\boxed{}\right)^2=16$$

$$\frac{1}{16}\left[4(x+3)^2+(y-1)^2\right]=\frac{1}{16}\cdot 16$$

$$\frac{(x+3)^2}{\boxed{}}+\frac{(y-1)^2}{\boxed{}}=\boxed{}$$

$$\frac{\left[x-(-3)\right]^2}{2^2}+\frac{(y-1)^2}{4^2}=1$$

General form: $\dfrac{(x-h)^2}{b^2} + \dfrac{(y-k)^2}{a^2} = 1$

Center: $\left(\boxed{}, 1\right)$

Vertices: $(-3, 1+4)$, $(-3, 1-4)$, or

$\left(-3, \boxed{}\right)$, $(-3, -3)$

Foci: $\left(-3, \boxed{} + 2\sqrt{3}\right)$, $\left(-3, 1-2\sqrt{3}\right)$

$c^2 = a^2 - b^2$

$c^2 = 16 - 4 = 12$

$c = \sqrt{12} = \boxed{}$

Endpoints of minor axis: $(-3+2, 1)$, $(-3-2, 1)$, or

$\left(\boxed{}, 1\right)$, $\left(\boxed{}, 1\right)$

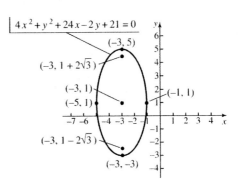

Section 10.3 The Hyperbola

Hyperbola

A **hyperbola** is the set of all points in a plane for which
the absolute value of the difference of the distances from
two fixed points (the **foci**) is constant. The midpoint of the
segment between the foci is the **center** of the hyperbola.

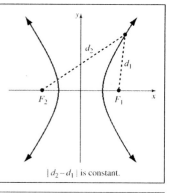

$|d_2 - d_1|$ is constant.

Standard Equation of a Hyperbola with Center at the Origin

Transverse Axis Horizontal

$$\frac{x^2}{a^2} - \frac{y^2}{b^2} = 1$$

Vertices: $(-a, 0)$, $(a, 0)$

Foci: $(-c, 0)$, $(c, 0)$, where $c^2 = a^2 + b^2$

Asymptotes: $y = \dfrac{b}{a}x$,

$$y = -\frac{b}{a}x$$

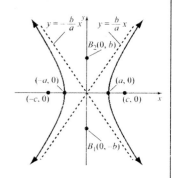

Transverse Axis Vertical

$$\frac{y^2}{a^2} - \frac{x^2}{b^2} = 1$$

Vertices: $(0, -a)$, $(0, a)$

Foci: $(0, -c)$, $(0, c)$, where $c^2 = a^2 + b^2$

Asymptotes: $y = \dfrac{a}{b}x$,

$$y = -\frac{a}{b}x$$

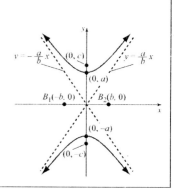

Example 1   Find an equation of the hyperbola with vertices $(0, -4)$ and $(0, 4)$ and
foci $(0, -6)$ and $(0, 6)$.

$a = 4, \quad c = 6$

$c^2 = a^2 + b^2$

$6^2 = \boxed{}^2 + b^2$

$36 = 16 + b^2$

$\boxed{} = b^2$

$$\frac{y^2}{a^2} - \frac{x^2}{b^2} = 1$$

$$\frac{y^2}{\boxed{}} - \frac{x^2}{\boxed{}} = 1$$

Example 2   For the hyperbola given by

$$9x^2 - 16y^2 = 144,$$

find the vertices the foci, and the asymptotes. Then graph the hyperbola.

$$\boxed{} \cdot \left(9x^2 - 16y^2\right) = \frac{1}{144} \cdot 144$$

$$\frac{x^2}{16} - \frac{y^2}{\boxed{}} = 1$$

$$\frac{x^2}{\boxed{}^2} - \frac{y^2}{3^2} = 1$$

$$a = 4, \qquad b = 3$$

Vertices: $\left(-4, 0\right),\ \left(\boxed{}, 0\right)$

$$c^2 = a^2 + b^2 = 16 + 9 = 25$$

$$c = \boxed{}$$

Foci: $\left(\boxed{}, 0\right),\ \left(5, 0\right)$

Asymptotes: $\quad y = -\dfrac{b}{a}x \qquad\qquad y = \dfrac{b}{a}x$

$$y = \boxed{}\,x \qquad\qquad y = \frac{3}{4}x$$

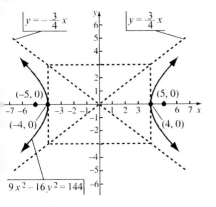

The Conics HYPERBOLA APP on most graphing calculaors can be used to graph hyperbolas when they are written in standard form.

Standard Equation of a Hyperbola with Center at (h, k)

Transverse Axis Horizontal

$$\frac{(x-h)^2}{a^2} - \frac{(y-k)^2}{b^2} = 1$$

Vertices: $(h-a, k)$, $(h+a, k)$

Asymptotes: $y - k = \dfrac{b}{a}(x-h)$, $y - k = -\dfrac{b}{a}(x-h)$

Foci: $(h-c, k)$, $(h+c, k)$, where $c^2 = a^2 + b^2$

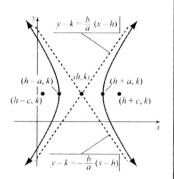

Transverse Axis Vertical

$$\frac{(y-k)^2}{a^2} - \frac{(x-h)^2}{b^2} = 1$$

Vertices: $(h, k-a)$, $(h, k+a)$

Asymptotes: $y - k = \dfrac{a}{b}(x-h)$, $y - k = -\dfrac{a}{b}(x-h)$

Foci: $(h, k-c)$, $(h, k+c)$, where $c^2 = a^2 + b^2$

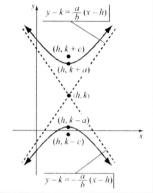

Example 3 For the hyperbola given by

$$4y^2 - x^2 + 24y + 4x + 28 = 0,$$

find the center, the vertices, the foci, and the asymptotes. Then draw the graph.

$$4y^2 - x^2 + 24y + 4x + 28 = 0$$

$$4\left(y^2 + 6y \right) - \left(x^2 - 4x \right) = -28$$

$$4\left(y^2 + 6y + \boxed{}\right) - \left(x^2 - 4x + \boxed{}\right) = -28$$

$$4\left(y^2 + 6y + 9\right) + 4\left(\boxed{}\right) - \left(x^2 - 4x + 4\right) + (-1)\left(\boxed{}\right) = -28$$

$$4\left(y^2 + 6y + 9\right) - \boxed{} - \left(x^2 - 4x + 4\right) + 4 = -28$$

$$4\left(\boxed{}\right)^2 - (x-2)^2 = \boxed{}$$

$$\frac{(y+3)^2}{1} - \frac{(x-2)^2}{\boxed{}} = 1$$

Standard form:

$$\frac{\left[y-(-3)\right]^2}{1^2} - \frac{(x-2)^2}{2^2} = 1 \qquad \frac{(y-k)^2}{a^2} - \frac{(x-h)^2}{b^2} = 1$$

Center: $\left(\boxed{}, -3\right)$

Vertices: $\left(2, -3-1\right)$, $\left(2, -3+1\right)$, or

$\qquad\left(2, -4\right)$, $\left(2, \boxed{}\right)$

Foci: $\left(2, -3-\sqrt{5}\right)$, $\left(2, -3+\sqrt{5}\right)$

$a^2 = 1, \qquad b^2 = 4$

$c^2 = a^2 + b^2$

$c^2 = 1 + 4$

$\quad= 5$

$c = \boxed{}$

Asymptotes:

$$y - k = \frac{a}{b}(x-h), \qquad\qquad y - k = -\frac{a}{b}(x-h)$$

$$y - (-3) = \boxed{}(x-2), \qquad y - (-3) = \boxed{}(x-2)$$

$$y + 3 = \frac{1}{2}(x-2), \qquad\qquad y + 3 = -\frac{1}{2}(x-2)$$

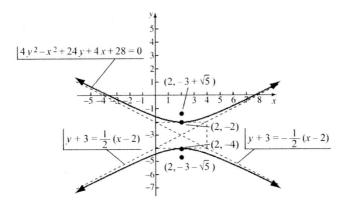

Section 10.4   Nonlinear Systems of Equations and Inequalities

Example 1   Solve the following system of equations:

$x^2 + y^2 = 25,$ (1)   The graph is a circle.

$3x - 4y = 0.$ (2)   The graph is a line.

$3x - 4y = 0$

$\qquad 3x = 4y$

$\qquad x = \boxed{}$

$x^2 + y^2 = 25$ (1)

$\left(\boxed{}\right)^2 + y^2 = 25$ $\qquad\qquad\qquad\qquad x = \dfrac{4}{3}y$

$\dfrac{16}{9}y^2 + y^2 = 25$ $\qquad\qquad\qquad x = \dfrac{4}{3}\cdot\boxed{} = 4 \quad \left(\boxed{}, 3\right)$

$\boxed{}\cdot y^2 = 25$ $\qquad\qquad\qquad x = \dfrac{4}{3}\cdot\left(\boxed{}\right) = -4 \quad \left(\boxed{}, -3\right)$

$\qquad y^2 = 9$

$\qquad y = \boxed{}$

Check for $(4, 3)$:

$$
\begin{array}{c|c}
\multicolumn{2}{c}{x^2 + y^2 = 25} \\
\hline
(4)^2 + \left(\boxed{}\right)^2 \stackrel{?}{\,} & 25 \\
16 + 9 & \\
25 & 25 \quad \text{TRUE}
\end{array}
\qquad
\begin{array}{c|c}
\multicolumn{2}{c}{3x - 4y = 0} \\
\hline
3\left(\boxed{}\right) - 4(3) \stackrel{?}{\,} & 0 \\
12 - 12 & \\
0 & 0 \quad \text{TRUE}
\end{array}
$$

Check for $(-4, -3)$:

$$
\begin{array}{c|c}
\multicolumn{2}{c}{x^2 + y^2 = 25} \\
\hline
\left(\boxed{}\right)^2 + (-3)^2 \stackrel{?}{\,} & 25 \\
16 + 9 & \\
25 & 25 \quad \text{TRUE}
\end{array}
\qquad
\begin{array}{c|c}
\multicolumn{2}{c}{3x - 4y = 0} \\
\hline
3(-4) - 4\left(\boxed{}\right) \stackrel{?}{\,} & 0 \\
-12 + 12 & \\
0 & 0 \quad \text{TRUE}
\end{array}
$$

The solutions are $(4, 3)$ and $(-4, -3)$.

Example 2 Solve the following system of equations

$x + y = 5$, (1) The graph is a line.

$y = 3 - x^2$. (2) The graph is a parabola.

$x + \boxed{} = 5$

$x - 2 - x^2 = 0$

$x^2 - x + 2 = 0 \qquad a = 1, \qquad b = \boxed{}, \qquad c = 2$

$x = \dfrac{-b \pm \sqrt{b^2 - 4ac}}{2a} = \dfrac{-(-1) \pm \sqrt{\left(\boxed{}\right)^2 - 4 \cdot 1 \cdot \boxed{}}}{2 \cdot 1}$

$= \dfrac{1 \pm \sqrt{1 - 8}}{2} = \dfrac{1 \pm \sqrt{\boxed{}}}{2}$

$= \dfrac{1 \pm \boxed{} \cdot \sqrt{7}}{2} = \dfrac{1}{2} \pm \dfrac{\sqrt{7}}{2} i$

$\boxed{} + y = 5$

$y = \dfrac{9}{2} - \dfrac{\sqrt{7}}{2} i$

$\boxed{} + y = 5$

$y = \dfrac{9}{2} + \dfrac{\sqrt{7}}{2} i$

The solutions are

$\left(\dfrac{1}{2} + \dfrac{\sqrt{7}}{2} i, \dfrac{9}{2} - \dfrac{\sqrt{7}}{2} i \right)$ and $\left(\dfrac{1}{2} - \dfrac{\sqrt{7}}{2} i, \dfrac{9}{2} + \dfrac{\sqrt{7}}{2} i \right)$.

Example 3   Solve the following system of equations:

$2x^2 + 5y^2 = 39,$ (1)   The graph is an ellipse.

$3x^2 - y^2 = -1,$ (2)   The graph is a hyperbola.

$$2x^2 + 5y^2 = 39$$

$$\underline{15x^2 - 5y^2 = \boxed{}}$$

$$17x^2 \qquad = 34$$

$$x^2 = 2$$

$$x = \boxed{}$$

$$3x^2 - y^2 = -1 \qquad\qquad (2)$$

$$3\left(\boxed{}\right)^2 - y^2 = -1 \qquad\qquad \text{The solutions are}$$

$$3 \cdot 2 - y^2 = -1 \qquad\qquad\qquad \left(\sqrt{2}, \boxed{}\right),$$

$$6 - y^2 = -1 \qquad\qquad\qquad \left(\sqrt{2}, \sqrt{7}\right),$$

$$-y^2 = -7 \qquad\qquad\qquad \left(\boxed{}, \sqrt{7}\right),$$

$$y^2 = 7 \qquad\qquad\qquad \left(-\sqrt{2}, -\sqrt{7}\right).$$

$$y = \boxed{}$$

Example 4   Solve the following system of equations:

$x^2 - 3y^2 = 6,$ (1)

$xy = 3.$ (2)

$$xy = 3$$

$$y = \frac{3}{\boxed{}} \qquad\qquad x^2 - 3y^2 = 6$$

$$x^2 - 3\left(\boxed{}\right)^2 = 6$$

$$x^2 - 3 \cdot \frac{9}{x^2} = 6$$

$$x^2 - \frac{27}{x^2} = 6$$

$$x^4 - 27 = \boxed{}$$

$$x^4 - 6x^2 - 27 = 0 \qquad u = \boxed{}$$

$$u^2 - 6 \cdot \boxed{} - 27 = 0$$

$$\left(\boxed{}\right)(u + 3) = 0$$

$$u - 9 = 0 \qquad \text{or} \quad u + 3 = 0$$

$$u = 9 \qquad \text{or} \qquad u = -3$$

$$x^2 = 9 \qquad \text{or} \qquad x^2 = -3$$

$$x = \boxed{} \quad \text{or} \qquad x = \boxed{}$$

$$y = \frac{3}{x}$$

The solutions are

$$x = 3; \ y = \frac{3}{3} = \boxed{}$$

$$(3, 1),$$

$$x = -3; \ y = \frac{3}{-3} = -1$$

$$\left(\boxed{}, -1\right),$$

$$x = i\sqrt{3}; \ y = \frac{3}{i\sqrt{3}}$$

$$\left(i\sqrt{3}, -i\sqrt{3}\right), \text{ and}$$

$$y = \frac{3}{i\sqrt{3}} \cdot \frac{-i\sqrt{3}}{-i\sqrt{3}} = \boxed{}$$

$$x = -i\sqrt{3}; \ y = \frac{3}{-i\sqrt{3}}$$

$$\left(-i\sqrt{3}, \boxed{}\right).$$

$$y = \frac{3}{-i\sqrt{3}} \cdot \frac{i\sqrt{3}}{i\sqrt{3}} = i\sqrt{3}$$

Example 5   *Dimensions of a Piece of Land.* For a student recreation building at Southport Community College, an architect wants to lay out a rectangular piece of land that has a perimeter of 204 m and an area of 2565 m². Find the dimensions of the piece of land.

1. Familiarize.

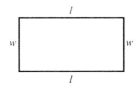

$l =$ length in meters

$w =$ width in meters

$P = 204$ m, $A = 2565$ m²

2. Translate.

$$2w + 2l = \boxed{}$$
$$lw = \boxed{}$$

3. Carry Out.

$$2w + 2l = 204$$
$$lw = 2565$$
$$l = \frac{2565}{\boxed{}}$$
$$2w + 2\left(\boxed{}\right) = 204$$
$$2w^2 + 2(2565) = 204w$$
$$2w^2 - 204w + 2(2565) = 0$$
$$w^2 - \boxed{}\,w + 2565 = 0$$
$$(w - 57)\left(\boxed{}\right) = 0$$
$$w - 57 = 0 \quad \text{or} \quad w - 45 = 0$$
$$w = \boxed{} \quad \text{or} \quad w = 45$$

Substituting for w in $l = \dfrac{2565}{w}$, we have

$$w = 57: \quad l = \frac{2565}{\boxed{}} = 45,$$

$$w = 45: \quad l = \frac{2565}{\boxed{}} = 57.$$

Considering l longer than w, we have $l = \boxed{}$ and $w = 45$.

4. **Check.**

   Perimeter is $2 \cdot \boxed{} + 2 \cdot 57 = \boxed{}$.

   Area is $57 \cdot 45 = \boxed{}$.

5. **State.**

   The length of the piece of land is 57 m, and the width is 45 m.

Example 6   Graph the solution set of the system

$$x^2 + y^2 \le 25,$$
$$3x - 4y > 0.$$

Related equations:

$$x^2 + y^2 \boxed{} 25 \qquad\qquad\qquad 3x - 4y \boxed{} 0$$

$$y = \frac{3x}{4}$$

Determine which region to shade.

Test $(0, 0)$: $\dfrac{x^2 + y^2 \le 25}{}$

$$0^2 + \boxed{}^2 \;?\; 25$$

$$0 \;\big|\; 25 \quad \text{TRUE}$$

Test $(0, 2)$: $\dfrac{3x - 4y > 0}{}$

$$3 \cdot 0 - 4 \cdot \boxed{} \;?\; 0$$

$$-8 \;\big|\; 0 \quad \text{FALSE}$$

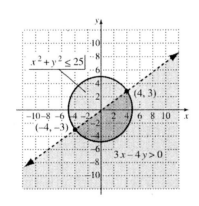

Example 7   Graph the solution set of the system

$$y \leq 4 - x^2,$$
$$x + y \geq 2.$$

Using a graphing calculator enter $y \boxed{} 4 - x^2$ with $\boxed{}$ as the relationship.

We also have

$$x + y \geq 2$$
$$y \geq \boxed{} - x.$$

Enter $y = 2 - x$ with $\boxed{}$ as the relationship.

If you don't have a calculator that allows you to enter the inequality symbols, use test points to determine whether to shade above or shade below each curve.

Test $(0, 0)$:   Test $(0, 0)$:

$$\frac{y \leq 4 - x^2}{0 \; \overset{?}{}\; 4 - 0^2}$$ $$\frac{x + y \geq 2}{0 + 0 \; \overset{?}{}\; 2}$$
$$0 \mid 4 \qquad \text{TRUE}$$ $$0 \mid 2 \qquad \text{FALSE}$$

Include the half-plane that contains   Exclude the half-plane that
$(0, 0)$ in the solution set.   contains $(0, 0)$.

We can use the INTERSECT feature to find the points of intersection of the graphs. They are $(-1, 3)$ and $\left(\boxed{}, 0\right)$.

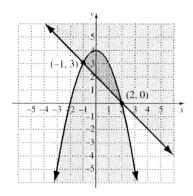

Section 10.5 Rotation of Axes

Rotation of Axes Formulas

If the x- and y-axes are rotated about the origin through a positive acute angle θ, then the coordinates (x, y) and (x', y') of a point P in the xy- and $x'y'$-coordinate systems are related by the following formulas:

$$x' = x\cos\theta + y\sin\theta, \qquad y' = -x\sin\theta + y\cos\theta,$$
$$x = x'\cos\theta - y'\sin\theta, \qquad y = x'\sin\theta + y'\cos\theta.$$

Example 1   Suppose that the xy-axes are rotated through an angle of $45°$. Write the equation $xy = 1$ in the $x'y'$-coordinate system.

We substitute $45°$ for $\boxed{}$ in the rotation of axes formulas.

$$x = x'\cos 45° - y'\sin 45°$$
$$y = x'\sin 45° + y'\cos 45°$$

$$\sin 45° = \frac{\sqrt{2}}{\boxed{}}, \quad \cos 45° = \frac{\boxed{}}{2}$$

$$x = x'\left(\frac{\sqrt{2}}{2}\right) - \boxed{}\left(\frac{\sqrt{2}}{2}\right) = \frac{\sqrt{2}}{2}(x' - y')$$

$$y = \boxed{}\left(\frac{\sqrt{2}}{2}\right) + y'\left(\frac{\sqrt{2}}{2}\right) = \frac{\sqrt{2}}{\boxed{}}(x' + y')$$

Substitute these expressions for x and y in the equation $xy = 1$.

$$\frac{\sqrt{2}}{2}(x' - y') \cdot \frac{\sqrt{2}}{2}(x' + y') = 1$$

$$\frac{1}{\boxed{}}\left[(x')^2 - (y')^2\right] = 1$$

$$\frac{(x')^2}{2} - \frac{(y')^2}{\boxed{}} = \boxed{}, \text{ or}$$

$$\frac{(x')^2}{\left(\sqrt{2}\right)^2} - \frac{(y')^2}{\left(\sqrt{2}\right)^2} = 1$$

This is the equation of a hyperbola in the $x'y'$-coordinate system with transverse axis on the $\boxed{}$-axis and vertices $\left(-\sqrt{2}, 0\right)$ and $\left(\sqrt{2}, \boxed{}\right)$. The asymptotes are $y' = x'$ and $y' = -\boxed{}$. These correspond to the axes of the xy-coordinate system.

Eliminating the xy-Term

To eliminate the xy-term from the equation

$$Ax^2 + Bxy + Cy^2 + Dx + Ey + F = 0, \ B \neq 0,$$

select an angle θ such that

$$\cot 2\theta = \frac{A - C}{B}, \ 0° < 2\theta < 180°,$$

and use the rotation of axes formulas.

Example 2   Graph the equation $3x^2 - 2\sqrt{3}xy + y^2 + 2x + 2\sqrt{3}y = 0$.

$A = 3, \ B = -2\sqrt{3}, \ C = \boxed{}, \ D = 2, \ E = 2\sqrt{3}, \ F = \boxed{}$

Now we find the angle of rotation.

$$\cot 2\theta = \frac{A - C}{B} = \frac{\boxed{} - \boxed{}}{-2\sqrt{3}} = \frac{2}{-2\sqrt{3}} = -\frac{1}{\sqrt{3}}$$

$$2\theta = 120°$$

$$\theta = \boxed{}°$$

Now substitute $60°$ for θ in the rotation of axes formulas.

$$x = x'\cos 60° - \boxed{}\sin 60°$$

$$y = \boxed{}\sin 60° + y'\cos 60°$$

$$x = x' \cdot \frac{1}{\boxed{}} - y' \cdot \frac{\sqrt{3}}{2} = \frac{x'}{2} - \frac{y'\sqrt{3}}{2}$$

$$y = x' \cdot \frac{\sqrt{3}}{2} + \boxed{} \cdot \frac{1}{2} = \frac{x'\sqrt{3}}{2} + \frac{y'}{\boxed{}}$$

Now we substitute these expressions for x and y in the given equation.

$$3\left(\frac{x'}{2} - \frac{y'\sqrt{3}}{2}\right)^2 - 2\sqrt{3}\left(\frac{x'}{2} - \frac{y'\sqrt{3}}{2}\right)\left(\frac{x'\sqrt{3}}{2} + \frac{y'}{2}\right) + \left(\frac{x'\sqrt{3}}{2} + \frac{y'}{2}\right)^2 +$$

$$\boxed{}\left(\frac{x'}{2} - \frac{y'\sqrt{3}}{2}\right) + 2\sqrt{3}\left(\frac{x'\sqrt{3}}{2} + \frac{y'}{2}\right) = 0$$

After simplifying, we get

$$4(y')^2 + 4x' = 0, \ \text{or} \ (y')^2 = -x'.$$

This is the equation of a parabola with vertex at $(0,0)$ in the $x'y'$-coordinate system and axis of symmetry $y' = \boxed{}$. Sketch the graph.

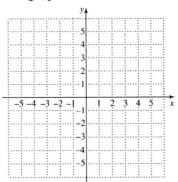

The graph of the equation

$$Ax^2 + Bxy + Cy^2 + Dx + Ey + F = 0$$

is, except in degenerate cases,

1. an ellipse or a circle if $B^2 - 4AC < 0$,
2. a hyperbola if $B^2 - 4AC > 0$, and
3. a parabola if $B^2 - 4AC = 0$.

The expression $B^2 - 4AC$ is the **discriminant** of the equation.

Example 3   Graph the equation $3x^2 + 2xy + 3y^2 = 16$.

$A = 3$, $B = \boxed{}$, $C = 3$

$B^2 - 4AC = 2^2 - 4 \cdot 3 \cdot \boxed{} = 4 - 36 = \boxed{}$

The discriminant is $\underline{\hspace{3cm}}$, so the graph is an ellipse or a circle. We find the
$\qquad\qquad$ positive / negative

angle of rotation.

$$\cot 2\theta = \frac{A - C}{B} = \frac{3 - \boxed{}}{2} = \frac{\boxed{}}{2} = \boxed{}$$
$$2\theta = 90°$$
$$\theta = 45°$$

$$\sin\theta = \frac{\sqrt{2}}{2}, \quad \cos\theta = \frac{\sqrt{2}}{2}$$

We substitute in the rotation of axes formulas.

$$x = x' \cdot \frac{\sqrt{2}}{2} - \boxed{} \cdot \frac{\sqrt{2}}{2} = \frac{\sqrt{2}}{2}(x' - y')$$

$$y = \boxed{} \cdot \frac{\sqrt{2}}{2} + y' \cdot \frac{\sqrt{2}}{2} = \frac{\sqrt{2}}{2}(x' + y')$$

Now we substitute these expressions for x and y in the given equation.

$$3\left[\frac{\sqrt{2}}{2}(x' - y')\right]^2 + \boxed{}\left[\frac{\sqrt{2}}{2}(x' - y')\right]\left[\frac{\sqrt{2}}{2}(\boxed{} + y')\right] + \boxed{}\left[\frac{\sqrt{2}}{2}(x' + y')\right]^2 = 16$$

After simplifying we have

$$4(x')^2 + 2(y')^2 = 16, \text{ or } \frac{(x')^2}{4} + \frac{(y')^2}{8} = \boxed{}.$$

This is the equation of an ellipse with vertices $\left(0, -\sqrt{8}\right)$ and $\left(\boxed{}, \sqrt{8}\right)$ on the y'-axis. The x'-intercepts are $(-2, 0)$ and $\left(\boxed{}, \boxed{}\right)$. Sketch the graph.

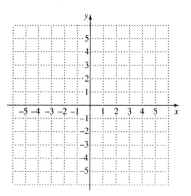

Example 4   Graph the equation $4x^2 - 24xy - 3y^2 - 156 = 0$.

$A = 4$, $B = -24$, $C = \boxed{}$

$B^2 - 4AC = (-24)^2 - 4 \cdot \boxed{}(-3) = 576 + 48 = \boxed{}$

The discriminant is _____ , so the graph is a hyperbola. We find the angle of
positive / negative

rotation.

$$\cot 2\theta = \frac{A - C}{B} = \frac{\boxed{} - (-3)}{-24} = -\frac{\boxed{}}{24}$$

Since $\cot 2\theta < 0$, we have $90° < 2\theta < 180°$. We make a drawing.

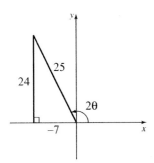

We see that $\cos 2\theta = -\dfrac{7}{\boxed{}}$. Now we use half-angle formulas to find $\sin\theta$ and $\cos\theta$.

$$\sin\theta = \sqrt{\frac{1-\cos 2\theta}{2}} = \sqrt{\frac{\boxed{}-\left(-\dfrac{7}{25}\right)}{2}} = \frac{4}{5}$$

$$\cos\theta = \sqrt{\frac{1+\cos 2\theta}{2}} = \sqrt{\frac{1+\left(-\dfrac{7}{25}\right)}{2}} = \frac{3}{5}$$

Now we substitute these expressions in the rotation of axes formulas.

$$x = x'\cos\theta - y'\sin\theta = \frac{3}{5}x' - \frac{4}{5}\boxed{}$$

$$y = x'\sin\theta + y'\cos\theta = \frac{4}{5}\boxed{} + \frac{3}{5}y'$$

Next we substitute for x and y in the given equation.

$$4\left(\frac{3}{5}x' - \frac{4}{5}y'\right)^2 - \boxed{}\left(\frac{3}{5}x' - \frac{4}{5}y'\right)\left(\frac{4}{5}x' + \frac{3}{5}y'\right) - 3\left(\frac{4}{5}x' + \frac{3}{5}y'\right)^2 - \boxed{} = 0$$

After simplifying, we get

$$13(y')^2 - 12(x')^2 - 156 = 0$$

$$13(y')^2 - 12(x')^2 = \boxed{}$$

$$\frac{(y')^2}{12} - \frac{(x')^2}{13} = \boxed{}.$$

The graph is a hyperbola with vertices $\left(0, -2\sqrt{3}\right)$ and $\left(\boxed{}, 2\sqrt{3}\right)$ on the y'-axis. Since

$\sin\theta = \boxed{}$, $\theta \approx 53.1°$, so the axes are rotated through an angle of about $\boxed{}°$. Sketch

the graph.

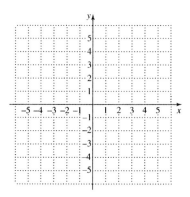

Section 10.6   Polar Equations of Conics

> **Polar Equations of Conics**
>
> A polar equation of any of the four forms
>
> $$r = \frac{ep}{1 \pm e\cos\theta}, \quad r = \frac{ep}{1 \pm e\sin\theta}$$
>
> is a conic section. The conic is a parabola if $e = 1$, an ellipse if $0 < e < 1$, and a hyperbola if $e > 1$.
>
> The number e is called the **eccentricity** of the conic section.
>
> For an ellipse and a hyperbola, the eccentricity is given by
>
> $$e = \frac{c}{a},$$
>
> where c is the distance from the center to a focus and a is the distance from the center to a vertex.

The table below describes the polar equations of conics with a focus at the pole and the directrix either perpendicular to or parallel to the polar axis.

Equation	Description
$r = \dfrac{ep}{1 + e\cos\theta}$	Vertical directrix p units to the right of the pole (or focus)
$r = \dfrac{ep}{1 - e\cos\theta}$	Vertical directrix p units to the left of the pole (or focus)
$r = \dfrac{ep}{1 + e\sin\theta}$	Horizontal directrix p units above the pole (or focus)
$r = \dfrac{ep}{1 - e\sin\theta}$	Horizontal directrix p units below the pole (or focus)

Example 1   Describe and graph the conic $r = \dfrac{18}{6 + 3\cos\theta}$.

$$r = \frac{18}{6 + 3\cos\theta}$$

$$r = \frac{3}{\boxed{} + 0.5\cos\theta} \qquad \text{Dividing numerator and denominator by 6}$$

$e = 0.5$, and $0 < 0.5 < 1$, so the graph is an ellipse.

For $ep = 3$ when $e = 0.5$, we have $p = \boxed{}$. Thus, the ellipse has a

$\underline{}$ directrix that lies 6 units to the $\underline{}$ of the pole.
horizontal / vertical right / left

We find the vertices by letting $\theta = 0$ and $\theta = \boxed{}$. They are $(2, 0)$ and $(6, \pi)$. The center of the ellipse is at the midpoint of the segment connecting the vertices, $(2, \pi)$. The length of the major axis is $\boxed{}$, so we have $2a = 8$, or $a = \boxed{}$.

$$e = \frac{c}{a}, \text{ so } 0.5 = \frac{c}{\boxed{}}, \text{ or } c = \boxed{}.$$

$b^2 = a^2 - c^2 = 4^2 - 2^2 = 16 - \boxed{} = 12$, so $b = \sqrt{12} = 2\sqrt{3}$.

This is the distance from the center to the point directly above (or below) it on the ellipse. Sketch the graph.

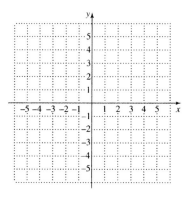

Example 2   Describe and graph the conic $r = \dfrac{10}{5 - 5\sin\theta}$.

$$r = \frac{10}{5 - 5\sin\theta}$$

$$r = \frac{2}{1 - \sin\theta} \qquad \begin{array}{l}\text{Dividing numerator and} \\ \text{denominator by 5}\end{array}$$

$e = 1$, so the graph is a parabola.

For $ep = 2$ when $e = 1$, we have $p = \boxed{}$. Thus, the parabola has a

$\underline{}$ directrix that lies 2 units $\underline{}$ the pole. It follows that the
horizontal / vertical   above / below

parabola has a vertical axis of symmetry. Since the directrix lies below the focus, or pole, the

parabola opens $\underline{}$. The vertex is the midpoint of the segment of the axis of
                up / down

symmetry from the focus to the directrix. We find it by letting $\theta = 3\pi / 2$. It is

$\left(\boxed{}, 3\pi / 2 \right)$. Sketch the graph.

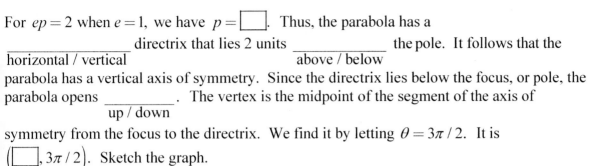

Example 3   Describe and graph the conic $r = \dfrac{4}{2 + 6\sin\theta}$.

$r = \dfrac{4}{2 + 6\sin\theta}$

$r = \dfrac{2}{1 + 3\sin\theta}$ Dividing numerator and
                                     denominator by 2

$e = 3$ and $3 > 1$, so the graph is a hyperbola.

For $ep = 2$ when $e = 3$, we have $p = \dfrac{2}{\boxed{}}$. Thus, the hyperbola has a $\underline{}$
                                                                                          horizontal / vertical

directrix that lies 2/3 unit $\underline{}$ the pole. Sketch the graph. Note that the vertices
                            above / below

are $\left(1/2, \pi / 2 \right)$ and $\left(-1, 3\pi / 2 \right)$, and the center is $\left(3/4, \pi / 2 \right)$.

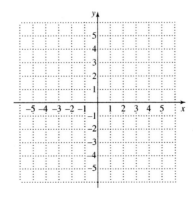

Example 4 Convert to a rectangular equation: $r = \dfrac{2}{1 - \sin\theta}$.

$$r = \dfrac{2}{1 - \sin\theta}$$

$$r - r\sin\theta = \boxed{}$$

$$r = r\sin\theta + 2$$

$$\sqrt{x^2 + y^2} = \boxed{} + 2$$

$$x^2 + y^2 = y^2 + 4y + \boxed{}$$

$$x^2 = 4y + 4, \text{ or}$$

$$x^2 - 4y - 4 = \boxed{}$$

This is the equation of a parabola as we should have anticipated since $e = \boxed{}$.

Example 5 Find a polar equation of the conic with a focus at the pole, eccentricity $\dfrac{1}{3}$, and directrix $r = 2\csc\theta$.

$$r = \dfrac{2}{\sin\theta}$$

$$r\sin\theta = \boxed{}$$

$$y = 2 \qquad \text{Rectangular equation of directrix}$$

We see that the directrix is a _____ line 2 units _____ the polar axis.
horizontal / vertical   above / below
Thus, the equation is of the form

$$r = \dfrac{ep}{1 + e\sin\theta}.$$

Substituting $\boxed{}$ for e and 2 for p gives us

$$r = \dfrac{\dfrac{1}{3} \cdot 2}{1 + \dfrac{1}{3}\sin\theta} = \dfrac{\dfrac{2}{3}}{\boxed{} + \dfrac{1}{3}\sin\theta}, \text{ or } \dfrac{2}{3 + \sin\theta}.$$

Section 10.7 Parametric Equations

Example 1   Graph the curve represented by the equations

$$x = \frac{1}{2}t, \ y = t^2 - 3; \ -3 \le t \le 3.$$

We choose values of t between -3 and $\boxed{}$ and find the corresponding values of x and y.
Let $t = -3$.

$$x = \frac{1}{2}(-3) = -\frac{3}{2}, \ y = (-3)^2 - 3 = \boxed{}$$

We list other ordered pairs in a table, plot those points, and draw the curve.

t	x	y	(x, y)
-3	$-\dfrac{3}{2}$	6	$\left(-\dfrac{3}{2}, 6\right)$
-2	-1	1	$(-1, 1)$
-1	$-\dfrac{1}{2}$	-2	$\left(-\dfrac{1}{2}, -2\right)$
0	0	-3	$(0, -3)$
1	$\dfrac{1}{2}$	-2	$\left(\dfrac{1}{2}, -2\right)$
2	1	1	$(1, 1)$
3	$\dfrac{3}{2}$	6	$\left(\dfrac{3}{2}, 6\right)$

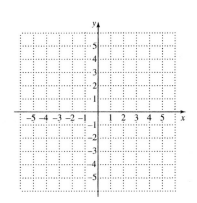

Parametric Equations

If f and g are continuous functions of t on an interval I, then the set of ordered pairs (x, y) such that $x = f(t)$ and $y = g(t)$ is a **plane curve**. The equations $x = f(t)$ and $y = g(t)$ are **parametric equations** for the curve. The variable t is the **parameter**.

Example 2 Using a graphing calculator, graph each of the following plane curves given their respective parametric equations and the restrictions on the parameter. Then find the equivalent rectangular equation.

a) $x = t^2$, $y = t - 1$; $-1 \le t \le 4$

b) $x = \sqrt{t}$, $y = 2t + 3$; $0 \le t \le 3$

a)   Set the calculator in PARAMETRIC mode (PAR).

Enter x and y. Then choose a window. A good choice is

Tmin = −1	Xmin = −2	Ymin = −4
Tmax = 4	Xmax = 18	Ymax = 4
Tstep = 1/10	Xscl = 1	Yscl = 1.

Finally, press GRAPH. (Sketch the graph in the space below.)

Now we convert the parametric equations to a rectangular equation.

$$y = t - 1$$

$$y + \boxed{} = t$$

Now substitute $y + 1$ for t in $x = t^2$: $x = (y + 1)^2$. For $-1 \le t \le 4$, we have

$0 \le x \le \boxed{}$ and $-2 \le y \le \boxed{}$.

b)   Set the calculator in PARAMETRIC mode (PAR).

Enter x and y, and then choose a window. A good choice is

Tmin = 0	Xmin = −3	Ymin = −2
Tmax = 3	Xmax = 3	Ymax = 10
Tstep = 1/10	Xscl = 1	Yscl = 1.

Finally, press GRAPH. (Sketch the graph in the space below.)

Now we convert the parametric equations to a rectangular equation.

$$x = \sqrt{t}$$
$$x^2 = t$$

Substitute x^2 for t is $y = 2t + 3$: $y = 2x^2 + \boxed{}$. For $0 \le t \le 3$, we have

$\boxed{} < x \le \sqrt{3}$ and $3 \le y \le \boxed{}$.

Example 3   Graph the plane curve represented by $x = \cos t$ and $y = \sin t$ with t in $[0, 2\pi]$. Then determine an equivalent rectangular equation.

Set the calculator in PARAMETRIC mode (PAR). Enter the equations and set up a squared window.

Tmin = 0	Xmin = −3	Ymin = −2
Tmax = 2π	Xmax = 3	Ymax = 2
Tstep = $\pi / 48$	Xscl = 1	Yscl = 1.

Finally, press GRAPH. The graph is the unit circle.

Now we find the equivalent rectangular equation.

$$x^2 = \cos^2 t, \; y^2 = \boxed{}$$

$$\cos^2 t + \sin^2 t = 1$$

$$x^2 + y^2 = \boxed{}$$

This is the rectangular equation of the unit circle.

Example 4   Graph the plane curve represented by

$$x = 5\cos t \text{ and } y = 3\sin t; \ 0 \le t \le 2\pi.$$

Then eliminate the parameter to find the rectangular equation.

Set the calculator in PARAMETRIC mode (PAR).

Enter the equations and set up a squared window.

Tmin $= 0$	Xmin $= -9$	Ymin $= -6$
Tmax $= 2\pi$	Xmax $= 9$	Ymax $= 6$
Tstep $= \pi / 48$	Xscl $= 1$	Yscl $= 1.$

Finally, press GRAPH. The graph is an ellipse centered at the origin with vertices at $(-5, 0)$ and $(5, 0)$.

Now we find the equivalent rectangular equation.

$$x = 5\cos t \qquad\qquad y = 3\sin t$$

$$\frac{x}{5} = \cos t \qquad\qquad \frac{y}{\boxed{}} = \sin t$$

$$\sin^2 t + \cos^2 t = 1$$

$$\left(\frac{y}{3}\right)^2 + \left(\frac{x}{\boxed{}}\right)^2 = 1$$

$$\frac{y^2}{\boxed{}} + \frac{x^2}{25} = 1, \text{ or } \frac{x^2}{25} + \frac{y^2}{9} = 1$$

Again, we see that we have an ellipse centered at the origin with vertices $(-5, 0)$ and $(5, 0)$.

Example 5   Using the VALUE feature from the CALC menu and the parametric graph of the unit circle, find each of the following function values.

a) $\cos\dfrac{7\pi}{6}$ b) $\sin 4.13$

a)   Set the calculator in PARAMETRIC mode (PAR) and also in RADIAN mode. Then enter the parametric equations of the unit circle, $x = \cos t$, $y = \sin t$, and set up a squared window.

Tmin = 0	Xmin = −3	Ymin = −2
Tmax = 2π	Xmax = 3	Ymax = 2
Tstep = $\pi / 48$	Xscl = 1	Yscl = 1.

Press GRAPH to see the unit circle. Now choose VALUE from the CALC menu, and enter $7\pi / 6$ for T. The x-value, −0.8660254, is $\cos t$. Thus, $\cos\dfrac{7\pi}{6} \approx -0.8660$.

b)   Graph the unit circle as described in part (a) above. Choose VALUE from the CALC menu, and enter 4.13 for t. The y-value, −0.835151, is $\sin t$. Thus, $\sin 4.13 \approx -0.8352$.

Example 6   Find three sets of parametric equations for the parabola $y = 4 - (x + 3)^2$.

If $x = t$, then $y = 4 - (t + 3)^2$, or $y = -t^2 - 6t - 5$.

If $x = t - 3$, then $y = 4 - \left(t - \boxed{} + 3\right)^2$, or $y = 4 - t^2$.

If $x = \dfrac{t}{3}$, then $y = \boxed{} - \left(\dfrac{t}{3} + \boxed{}\right)^2$, or $y = -\dfrac{t^2}{9} - 2t - 5$.

There are infinitely many such equations.

The motion of an object propelled upward, or **projectile motion**, can be described with parametric equations. For an object propelled upward at an angle θ with the horizontal from a height h, in feet, at an initial speed v_0, in feet per second, and neglecting air resistance, we have

$$x = (v_0 \cos\theta)t \text{ and } y = h + (v_0 \sin\theta)t - 16t^2.$$

Example 7   A baseball is thrown from a height of 6 ft with an initial speed of 100 ft/sec at an angle of $45°$ with the horizontal.

a)   Find parametric equations that give the position of the ball at time t, in seconds.

b)   Graph the plane curve represented by the equations found in part (a).

c)   Find the height of the ball after 1 sec, 2 sec, and 3 sec.

d)   Determine how long the ball is in the air.

e)   Determine the horizontal distance that the ball travels.

f)   Find the maximum height of the ball.

a) $x = \left(v_0 \cos\theta\right)t = \left(100\cos 45°\right)t$

$x = \left(100 \cdot \dfrac{\sqrt{2}}{2}\right)t = \boxed{}\sqrt{2}\,t$

$y = h + \left(v_0 \sin\theta\right)t - 16t^2$

$y = 6 + \left(\boxed{}\sin 45°\right)t - 16t^2$

$y = \boxed{} + \left(100 \cdot \dfrac{\sqrt{2}}{2}\right)t - 16t^2$

$y = 6 + \boxed{}\sqrt{2}\,t - 16t^2$

b)   Set the calculator in PARAMETRIC mode (PAR). Enter the equations found in part (a) and set up a window.

Tmin $= 0$	Xmin $= 0$	Ymin $= 0$
Tmax $= 7$	Xmax $= 350$	Ymax $= 100$
Tstep $= 1/10$	Xscl $= 50$	Yscl $= 10$

Finally, press GRAPH. (Sketch the graph in the space below.)

c) The y-value represents the height at time t. We can use a table on a graphing calculator, set in ASK mode, to find the desired y-values.

We can also find these values by substituting in the equation for y.

For $t = 1$, $y = 6 + 50\sqrt{2} \cdot \boxed{} - 16 \cdot 1^2 \approx 60.7$ ft

For $t = 2$, $y = \boxed{} + 50\sqrt{2} \cdot 2 - \boxed{} \cdot 2^2 \approx 83.4$ ft

For $t = 3$, $y = 6 + \boxed{}\sqrt{2} \cdot 3 - 16 \cdot 3^2 \approx 74.1$ ft

d) The ball hits the ground when $y = 0$.

$$y = 6 + 50\sqrt{2}\,t - 16t^2$$
$$\boxed{} = 6 + 50\sqrt{2}\,t - 16t^2, \text{ or}$$
$$-16t^2 + 50\sqrt{2}\,t + 6 = 0$$

This is a quadratic equation with $a = \boxed{}$, $b = 50\sqrt{2}$, and $c = \boxed{}$.

$$t = \frac{-50\sqrt{2} \pm \sqrt{\left(50\sqrt{2}\right)^2 - 4\left(\boxed{}\right)(6)}}{2\left(\boxed{}\right)}$$

$t \approx -0.1$ *or* $t \approx 4.5$

The negative value of t has no meaning in this application because t represents time. Thus, the ball is in the air for about $\boxed{}$ sec.

e) The ball is in the air for about 4.5 sec, so we have

$$x = 50\sqrt{2}\,t \approx 50\sqrt{2}\left(\boxed{}\right) \approx 318.2 \text{ ft.}$$

f) The maximum height of the ball occurs at the vertex of the function $y = 6 + 50\sqrt{2}\,t - 16t^2$.

$$t = -\frac{b}{2a} = -\frac{50\sqrt{2}}{2\left(\boxed{}\right)} \approx 2.2$$

For $t \approx 2.2$, we have

$$y = \boxed{} + 50\sqrt{2}\left(\boxed{}\right) - 16(2.2)^2 \approx 84.1 \text{ ft.}$$

Section 11.1 Sequences and Series

> **Sequences**
>
> - An **infinite sequence** is a function having for its domain the set of positive integers, $\{1, 2, 3, 4, 5, \ldots\}$.
>
> - An **finite sequence** is a function having for its domain a set of positive integers, $\{1, 2, 3, 4, 5, \ldots, n\}$, for some positive integer n.
>
> - The first term of a sequence is denoted as a_1, the fifth term as a_5, and the nth term, or **general term**, as a_n.

Example 1   Find the first 4 terms and the 23rd term of the sequence whose general term is given by $a_n = (-1)^n n^2$.

$$a_1 = (-1)^1 \cdot 1^2 = -1 \cdot 1 = \boxed{}$$

$$a_2 = (-1)^2 \cdot 2^2 = 1 \cdot 4 = \boxed{}$$

$$a_3 = (-1)^{\boxed{}} \cdot \boxed{}^2 = \boxed{} \cdot 9 = -9$$

$$a_4 = (-1)^{\boxed{}} \cdot \boxed{}^2 = \boxed{} \cdot 16 = 16$$

$$a_{23} = (-1)^{23} \cdot 23^2 = -1 \cdot 529 = \boxed{}$$

Example 2   Use a graphing calculator to find the first 5 terms of the sequence whose general term is given by $a_n = n/(n+1)$.

We can use a table or the SEQ feature. We will use the SEQ feature. Select SEQ from the LIST OPS menu and enter the general term as $X/(X+1)$, the variable x, the start and stop numbers 1 and $\boxed{}$ and a step value of 1. Press ENTER and also use the Frac feature to show the terms in fraction form. They are $\dfrac{1}{2}$, $\boxed{}$, $\dfrac{3}{4}$, $\boxed{}$, and $\dfrac{5}{6}$.

Example 3   Use a graphing calculator to graph the sequence whose general term is given by $a_n = n/(n+1)$.

Set the calculator in SEQ and DOT modes. Press $\boxed{Y =}$ and enter $u(n) = n/(n+1)$. We choose the window $[0, 10, 0, 1]$ with nMin $= 1$, nMax $= \boxed{}$, PlotStart $= 1$, and PlotStop $= 1$. Press GRAPH to see the graph. (Sketch the graph in the space below.)

Example 4   For each of the following sequences, predict the general term.

a) $1, \sqrt{2}, \sqrt{3}, 2, \ldots$

$\sqrt{1}, \sqrt{2}, \sqrt{3}, \boxed{}, \ldots$

$\sqrt{n}$

b) $-1, 3, -9, 27, -81, \ldots$

$1, 3, 9, \boxed{}, 81, \ldots$ $\qquad$ 3^{n-1}

c) $2, 4, 8, \ldots$

$2, 4, 8, 16, 32, \ldots$ $\qquad$ $2^{\boxed{}}$

$2, 4, 8, 14, 22, \ldots$ $\qquad$ $n^2 - n + 2$

Series

Given the infinite sequence

$$a_1, a_2, a_3, a_4, \ldots, a_n, \ldots,$$

the sum of the terms

$$a_1 + a_2 + a_3 + \cdots + a_n + \cdots$$

is called an **infinite series**. A **partial sum** is the sum of the first n terms:

$$a_1 + a_2 + a_3 + \cdots + a_n.$$

A partial sum is also called a **finite series**, or **nth partial sum**, and is denoted S_n.

Example 5   For the sequence $-2, 4, -6, 8, -10, 12, -14, \ldots,$ find each of the following sums.

a) $S_1 = \boxed{}$

b) $S_4 = -2 + 4 + \left(\boxed{}\right) + 8 = \boxed{}$

c) $S_5 = -2 + \boxed{} + (-6) + 8 + \left(\boxed{}\right) = \boxed{}$

Example 6   Use a graphing calculator to find S_1, S_2, S_3, and S_4 for the sequence whose general term is given by $a_n = n^2 - 3$.

We select the CUMSUM feature from the LIST OPS menu. Then go to LIST OPS again and choose Seq. Enter the expression $n^2 - 3$, the variable $\boxed{}$, the start and stop numbers $\boxed{}$ and 4, and a step value of 1. Finally, press ENTER twice to see the values -2, $\boxed{}$, 5, and $\boxed{}$ for S_1, S_2, S_3, and S_4, respectively.

Example 7   Find and evaluate each of the following sums.

a) $\displaystyle\sum_{k=1}^{5} k^3 = 1^3 + \boxed{}^3 + 3^3 + 4^3 + \boxed{}^3$

$\qquad = 1 + 8 + \boxed{} + \boxed{} + 125$

$\qquad = \boxed{}$

b) $\displaystyle\sum_{k=0}^{4} (-1)^k 5^k = (-1)^0 \cdot 5^0 + (-1)^1 \cdot 5^1 + (-1)^{\boxed{}} \cdot 5^{\boxed{}} + (-1)^3 \cdot 5^3 + (-1)^4 \cdot 5^4$

$\qquad = 1 \cdot \boxed{} + (-1) \cdot 5 + (1) \cdot \boxed{} + \left(\boxed{}\right) \cdot 125 + (1) \cdot 625$

$\qquad = 1 - 5 + 25 - \boxed{} + 625$

$\qquad = \boxed{}$

c) $\displaystyle\sum_{i=8}^{11} \left(2 + \frac{1}{i}\right) = \left(2 + \frac{1}{\boxed{}}\right) + \left(2 + \frac{1}{\boxed{}}\right) + \left(2 + \frac{1}{10}\right) + \left(2 + \frac{1}{11}\right)$

$\qquad = \boxed{} + \frac{1}{8} + \frac{1}{9} + \frac{1}{10} + \frac{1}{11}$

$\qquad = 8\dfrac{\boxed{}}{3960}$

Example 8   Write sigma notation for each sum.

a) $1+2+4+8+16+32+64$

$$= 2^{\square} + 2^1 + \boxed{}^2 + 2^3 + 2^{\square} + 2^5 + 2^6$$

$$= \sum_{k=0}^{\square} 2^{\square}$$

b) $-2+4-6+8-10$

$$= (-1)(2)+(1)(4)+\left(\boxed{}\right)(6)+(1)(8)+(-1)(10)$$

$$= (-1)\left(2\cdot\boxed{}\right)+(1)\left(2\cdot\boxed{}\right)+(-1)(2\cdot3)+(1)(2\cdot4)+(-1)(2\cdot5)$$

$$= (-1)^1 (2\cdot1)+\left(\boxed{}\right)^2 (2\cdot2)+(-1)^{\square} (2\cdot3)+(-1)^4 (2\cdot4)+(-1)^5 (2\cdot5)$$

$$= \sum_{k=1}^{5} \left(\boxed{}\right)^k \left(2\cdot\boxed{}\right)$$

c) $x+\dfrac{x^2}{2}+\dfrac{x^3}{3}+\dfrac{x^4}{4}+\cdots$

$$= \frac{x^1}{\boxed{}}+\frac{x^2}{2}+\frac{x^3}{3}+\frac{x^4}{4}+\cdots$$

$$= \sum_{k=1}^{\square} \frac{x^k}{\boxed{}}$$

Example 9   Find the first 5 terms of the sequence defined by

$$a_1 = 5, \quad a_{n+1} = 2a_n - 3, \quad \text{for } n \ge 1.$$

$a_1 = 5$

$a_2 = 2\cdot\boxed{} - 3 = 7$

$a_3 = 2\cdot7 - 3 = \boxed{}$

$a_4 = 2\cdot\boxed{} - 3 = 19$

$a_5 = 2\cdot19 - 3 = \boxed{}$

Section 11.2   Arithmetic Sequences and Series

Arithmetic Sequence

A sequence is **arithmetic** if there exists a number d, called the **common difference**, such that $a_{n+1} = a_n + d$ for any integer $n \geq 1$.

Example 1   For each of the following arithmetic sequences, identify the first term, a_1, and the common difference, d.

a) $4, 9, 14, 19, 24, \ldots$

First term: ☐

Common difference $d = 9 - 4 = $ ☐

$$14 - 9 = 5$$
$$19 - 14 = 5$$

b) $34, 27, 20, 13, 6, -1, -8, \ldots$

First term: ☐

$d = $ ☐ $- 34 = -7$

$$20 - 27 = -7$$
$$13 - 20 = -7$$

c) $2, 2\frac{1}{2}, 3, 3\frac{1}{2}, 4, 4\frac{1}{2}, \ldots$

First term: ☐

$d = 2\frac{1}{2} - 2 = $ ☐

$$3 - 2\frac{1}{2} = \frac{1}{2}$$
$$3\frac{1}{2} - 3 = \frac{1}{2}$$

nth Term of an Arithmetic Sequence

The **nth term** of an arithmetic sequence is given by

$$a_n = a_1 + (n-1)d, \text{ for any integer } n \geq 1.$$

Example 2   Find the 14th term of the arithmetic sequence $4, 7, 10, 13, \ldots$.

$a_n = a_1 + (n-1)d$ $a_1 = 4$

$a_{14} = $ ☐ $+ (14-1) \cdot$ ☐ $d = 7 - 4 = 3$

$\quad = 4 + 13 \cdot 3$ $n = 14$

$\quad = 4 + 39$

$\quad = 43$

The 14th term is ☐.

Example 3   In the sequence of $4, 7, 10, 13, \ldots,$ which term is 301? That is, find n if $a_n = 301$.

$$a_n = a_1 + (n-1)d$$

$$\boxed{} = 4 + (n-1) \cdot 3$$

$$301 = 4 + 3n - \boxed{}$$

$$301 = 3n + 1$$

$$300 = 3n$$

$$100 = n$$

$a_1 = 4$

$d = 7 - 4 = 3$

$a_n = 301$

$n = ?$

The term 301 is the $\boxed{}^{\text{th}}$ term.

Example 4   The 3rd term of an arithmetic sequence is 8, and the 16th term is 47. Find a_1 and d and construct the sequence.

$$8 + \boxed{} = 47$$

$$13d = 39$$

$$d = 3$$

$$a_1 = \boxed{} - 2 \cdot 3 = \boxed{}$$

$$2, \boxed{}, 8, 11, \ldots$$

Sum of the First n Terms

The sum of the first n terms of an arithmetic sequence is given by

$$S_n = \frac{n}{2}(a_1 + a_n).$$

Example 5   Find the sum of the first 100 natural numbers.

$$1 + 2 + 3 + \cdots + 99 + 100$$

$$S_n = \frac{n}{2}(a_1 + a_n)$$

$$S_{100} = \frac{\boxed{}}{2}(1 + 100) = \boxed{}$$

$a_1 = 1$

$a_n = \boxed{}$

$n = 100$

Example 6   Find the sum of the first 15 terms of the arithmetic sequence $4, 7, 10, 13, \ldots$.

$$a_1 = 4 \quad d = 3 \quad n = \boxed{}$$

$$S_n = \frac{n}{2}(a_1 + a_n)$$

$$S_{15} = \frac{\boxed{}}{2}(4 + 46)$$

$$= \frac{15}{2}(50)$$

$$= \boxed{}$$

$$a_n = a_1 + (n-1)d$$

$$a_{15} = \boxed{} + (15-1)3$$

$$= 4 + 14 \cdot 3 = \boxed{}$$

Example 7   Find the sum: $\displaystyle\sum_{k=1}^{130}(4k+5)$.

$$9 + \boxed{} + 17 + \cdots$$

$$a_1 = 9 \quad d = \boxed{} \quad n = 130$$

$$a_{130} = 4 \cdot \boxed{} + 5$$

$$a_{130} = 525$$

$$S_{130} = \frac{130}{2}\left(\boxed{} + 525\right)$$

$$S_{130} = \boxed{}$$

Example 8   Rachel accepts a job, starting with an hourly wage of $14.25, and is promised a raise of 15¢ per hour every 2 months for 5 years. At the end of 5 years, what will Rachel's hourly wage be?

$14.25 $14.25, 14.40, 14.55, \ldots$ $a_1 = 14.25$

$14.40 $a_n = a_1 + (n-1)d$ $d = 0.15$

$14.55

.   1 year $\dfrac{12}{2} = \boxed{}$ $n = \boxed{}$

.

.   5 years $5 \cdot 6 = \boxed{}$

$$a_{31} = \boxed{} + (31-1)0.15$$

$$a_{31} = 18.75$$

At the end of 5 years, Rachel's hourly wage will be $\boxed{}$.

Example 9   A stack of electric poles has 30 poles in the bottom row. There are 29 poles in the second row, 28 in the next row, and so on. How many poles are in the stack if there are 5 poles in the top row?

$$S_n = \frac{n}{2}(a_1 + a_n)$$

$$S_{26} = \frac{\boxed{}}{2}(30 + 5)$$

$$= 13(35)$$

$$= 455$$

$n = \boxed{}$

$a_1 = 30$

$a_{26} = \boxed{}$

There are $\boxed{}$ poles in the stack.

Section 11.3   Geometric Sequences and Series

Geometric Sequence

A sequence is **geometric** if there is a number r, called the **common ratio**, such that

$$\frac{a_{n+1}}{a_n} = r, \text{ or } a_{n+1} = a_n r, \text{ for any integer } n \geq 1.$$

Example 1   For each of the following geometric sequences, identify the common ratio.

a) $3, 6, 12, 24, 48, \ldots$

$r = \dfrac{6}{\boxed{}} = 2$

b) $1, -\dfrac{1}{2}, \dfrac{1}{4}, -\dfrac{1}{8}, \ldots$

$r = \dfrac{-\dfrac{1}{2}}{1} = \boxed{}$

c) $\$5200, \$3900, \$2925, \$2193.75, \ldots$

$r = \dfrac{\boxed{}}{5200} = 0.75$

d) $\$1000, \$1060, \$1123.60, \ldots$

$r = \dfrac{1060}{1000} = \boxed{}$

---

nth Term of a Geometric Sequence

The **nth term** of a geometric sequence is given by

$$a_n = a_1 r^{n-1}, \text{ for any integer } n \geq 1.$$

Example 2   Find the 7th term of the geometric sequence $4, 20, 100, \ldots$.

$a_n = a_1 r^{n-1}, \; n \geq 1$

$a_1 = 4 \quad n = \boxed{}, \quad r = \dfrac{20}{4} = \boxed{}$

$a_7 = \boxed{} \cdot 5^{7-1} = 4 \cdot 5^6$

$\qquad = 4 \cdot 15,625$

$\qquad = \boxed{}$

The 7th term is 62,500.

Example 3   Find the 10th term of the geometric sequence $64, -32, 16, -8, \ldots$.

$$a_n = a_1 r^{n-1},\ n \geq 1$$

$$a_1 = \boxed{}, \qquad n = 10, \qquad r = \frac{\boxed{}}{64} = -\frac{1}{2}$$

$$a_{10} = 64 \cdot \left(\boxed{}\right)^{10-1} = 64 \cdot \left(-\frac{1}{2}\right)^9$$

$$= 64 \cdot \left(-\frac{1}{2^9}\right)$$

$$= \boxed{} \cdot \left(-\frac{1}{2^9}\right)$$

$$= -\frac{1}{2^3} = \boxed{}$$

The 10th term is $-\dfrac{1}{8}$.

Sum of the First n Terms

The sum of the first n terms of a geometric sequence is given by

$$S_n = \frac{a_1\left(1 - r^n\right)}{1 - r}, \text{ for any } r \neq 1.$$

Example 4   Find the sum of the first 7 terms of the geometric sequence $3, 15, 75, 375, \ldots$.

$$a_1 = 3, \qquad n = 7, \qquad r = \frac{15}{\boxed{}} = 5$$

$$S_n = \frac{a_1\left(1 - r^n\right)}{1 - r},\ r \neq 1$$

$$S_7 = \frac{3\left(1 - 5^{\boxed{}}\right)}{1 - \boxed{}} = \frac{3\left(1 - 78{,}125\right)}{-4}$$

$$= \boxed{}$$

The sum of the first 7 terms is 58,593.

Example 5   Find the sum: $\displaystyle\sum_{k=1}^{11}(0.3)^k$.

$$S_n = \frac{a_1\left(1-r^n\right)}{1-r}$$

$a_1 = 0.3$

$r = \boxed{}$

$$S_{11} = \frac{0.3\left(1-0.3^{\boxed{}}\right)}{1-0.3}$$

$n = 11$

$$S_{11} \approx \boxed{}$$

Limit or Sum of an Infinite Geometric Series

When $|r| < 1$, the limit or sum of an infinite geometric series is given by

$$S_\infty = \frac{a_1}{1-r}.$$

Example 6   Determine whether each of the following infinite geometric series has a limit. If a limit exists, find it.

a) $1+3+9+27+\cdots$

$r = \dfrac{3}{1} = 3$

An infinite geometric series has a limit if $|r| < 1$.

$|r| = |3| = \boxed{}$

Since 3 is not less than $\boxed{}$, this series _____ have a limit.
does / does not

b) $-2+1-\dfrac{1}{2}+\dfrac{1}{4}-\dfrac{1}{8}+\cdots$

$r = \dfrac{1}{-2} = \boxed{}$

An infinite geometric series has a limit if $|r| < 1$.

$|r| = \left|-\dfrac{1}{2}\right| = \dfrac{1}{2}$

Since $\boxed{} < 1$, this series has a limit.

$$S_\infty = \frac{a_1}{1-r} = \frac{-2}{1-\boxed{}} = \frac{-2}{\frac{3}{2}} = -\frac{4}{3}$$

The limit of this series is $\boxed{}$.

Example 7   Find fraction notation for $0.78787878\ldots$, or $0.\overline{78}$.

$0.78 + 0.0078 + 0.000078 + \cdots$

$a_1 = 0.78 \quad r = \boxed{}$

$|r| = |0.01| = 0.01 < \boxed{}$

This infinite geometric series has a limit.

$$S_\infty = \frac{a_1}{1-r} = \frac{0.78}{1-\boxed{}} = \frac{0.78}{0.99} = \frac{78}{99} = \frac{\boxed{}}{33}$$

Example 8   Suppose someone offered you a job for the month of September (30 days) under the following conditions. You will be paid $0.01 for the first day, $0.02 for the second, $0.04 for the third, and so on, doubling your previous day's salary each day. How much would you earn altogether for the month? (Would you take the job? Make a conjecture before reading further.)

$0.01 \quad 0.02 \quad \boxed{} \ldots$

$0.01 \quad 0.01(2) \quad 0.01(2)(2) \ldots$

$0.01 + 0.01 \cdot 2^1 + 0.01 \cdot 2^{\boxed{}} + 0.01 \cdot 2^3 + \cdots$

$a_1 = 0.01 \quad r = \boxed{} \quad n = \boxed{}$

$$S_n = \frac{a_1\left(1 - r^n\right)}{1-r}$$

$$S_{30} = \frac{0.01\left(1 - 2^{30}\right)}{1 - \boxed{}} = \$10,737,418.23$$

Your pay exceeds $\boxed{}$ million for the month.

Example 9   An **annuity** is a sequence of equal payments, made at equal time intervals, that earns interest. Fixed deposits in a savings account are an example of an annuity. Suppose that to save money to buy a car, Jacob deposits $2000 at the *end* of each of 5 years in an account that pays 3% interest, compounded annually. The total amount in the account at the end of 5 years is called the **amount of the annuity**. Find that amount.

End of Year	Amount of Deposit		Amount Grows To
0	0		0
1	2000	$\rightarrow$	$2000\left(\boxed{}\right)^4$ (at 3% per year)
2	2000	$\rightarrow$	$2000(1.03)^{\square}$
3	2000	$\rightarrow$	$2000(1.03)^2$
4	2000	$\rightarrow$	$2000(1.03)^1$
5	2000	$\rightarrow$	2000

$$2000 + 2000(1.03)^1 + 2000(1.03)^2 + 2000(1.03)^3 + 2000(1.03)^4$$

$$a_1 = 2000 \qquad\qquad S_n = \frac{a_1\left(1-r^n\right)}{1-r}$$

$$r = \boxed{}$$

$$n = 5 \qquad\qquad S_5 = \frac{\boxed{}\left(1-1.03^5\right)}{1-1.03}$$

$$\approx \boxed{}$$

Jacob has been able to save $10,618.27.

Example 10   Large sporting events have a significant impact on the economy of the host city. Super Bowl XLVII, hosted by New Orleans, generated a $480-million net impact for the region. Assume that 60% of that amount is spent again in the area, and then 60% of that amount is spent again, and so on. This is known as the *economic multiplier effect*. Find the total effect on the economy.

$$480,000,000 + 480,000,000(0.60) + 480,000,000(0.60)^2 + \cdots$$

$$|r| < 1 \text{ because } r = \boxed{}$$

$$S_\infty = \frac{a_1}{1-r} = \frac{480,000,000}{1-0.6} = \frac{480,000,000}{\boxed{}}$$

$$= 1,200,000,000$$

$$= \boxed{} \text{ billion}$$

Section 11.4   Mathematical Induction

> **The Principle of Mathematical Induction**
>
> We can prove an infinite sequence of statements S_n by showing the following.
>
> (1)   *Basis step.* S_1 is true.
> (2)   *Induction step.* For all natural numbers k, $S_k \rightarrow S_{k+1}$.

Example 1   Prove. For every natural number n,

$$1+3+5+\cdots+(2n-1)=n^2.$$

$$S_n : 1+3+5+\cdots+(2n-1)=n^2$$

For $n=1$

$$S_1 : 1 = \boxed{}^2$$

For $n=k$

$$S_k : 1+3+5+\cdots+\left(\boxed{}\right)=\boxed{}^2$$

For $n=k+1$

$$S_{k+1} : 1+3+5+\cdots+(2k-1)+\left[2\left(\boxed{}\right)-1\right]=\left(\boxed{}\right)^2$$

1.   Basis step: S_1 is true.

2.   Induction step: Assume $\boxed{}$ is true.

$$\underbrace{1+3+5+\cdots+(2k-1)}+\left[2(k+1)-1\right]$$

$$=\boxed{}+\left[2(k+1)-1\right]$$
$$=k^2+2k+2-1$$
$$=k^2+2k+1$$
$$=\left(\boxed{}\right)^2$$

We have shown that for all natural numbers k, $S_k \rightarrow S_{k+1}$. This completes the induction step. It and the Basis Step tell us that the proof is complete.

Example 2 Prove: For every natural number n,

$$\frac{1}{2}+\frac{1}{4}+\frac{1}{8}+\cdots+\frac{1}{2^n}=\frac{2^n-1}{2^n}.$$

$S_n:\ \dfrac{1}{2}+\dfrac{1}{4}+\dfrac{1}{8}+\cdots+\dfrac{1}{2^n}=\dfrac{2^n-1}{2^n}$

$S_1:\ \dfrac{1}{2^{\square}}=\dfrac{2^{\square}-1}{2^{\square}}$

$S_k:\ \dfrac{1}{2}+\dfrac{1}{4}+\dfrac{1}{8}+\cdots+\dfrac{1}{2^{\square}}=\dfrac{2^{\square}-1}{2^{\square}}$

$S_{k+1}:\ \dfrac{1}{2}+\dfrac{1}{4}+\dfrac{1}{8}+\cdots+\dfrac{1}{2^k}+\dfrac{1}{2^{\square}}=\dfrac{2^{\square}-1}{2^{\square}}$

1. Basis step. S_1 is true.

2. Induction step: Assume S_k is true.

$$\underbrace{\frac{1}{2}+\frac{1}{4}+\frac{1}{8}+\cdots+\frac{1}{2^k}}+\frac{1}{2^{k+1}}$$

$$=\boxed{}+\frac{1}{2^{k+1}}=\frac{2^k-1}{2^k}\cdot\boxed{}+\frac{1}{2^{k+1}}$$

$$=\frac{2\left(2^k-1\right)}{2^{k+1}}+\frac{1}{2^{k+1}}=\frac{2^{k+1}-2+1}{2^{k+1}}$$

$$=\frac{2^{k+1}-1}{2^{k+1}}$$

We have shown that for all natural numbers k, $S_k \rightarrow S_{k+1}$. This completes the Induction Step. It and the Basis Step tell us that the proof is complete.

Example 3   Prove: For every natural number n, $n < 2^n$.

$S_n : n < 2^n$

$S_1 : 1 < 2^{\square}$

$S_k : \boxed{} < 2^k$

$S_{k+1} : k+1 < 2^{\boxed{}}$

1. Basis step: S_1 is true.

2. Induction step: Assume S_k is true.

$\quad k < 2^k \qquad$ This is S_k.

$\quad 2k < 2 \cdot 2^k \qquad$ Multiplying by 2 on both sides

$\quad 2k < 2^{\boxed{}} \qquad$ Adding exponents on the right

$\quad k + k < 2^{k+1} \qquad$ Rewriting $2k$ as $k + k$

Since k is any natural number, we know that $1 \le \boxed{}$.

$\quad k + 1 \le k + k \qquad$ Adding k on both sides of $1 \le k$.

Putting $k + k < 2^{k+1}$ and $k + 1 \le k + k$ together, we have

$\quad k + 1 \le \boxed{} < 2^{k+1}$

$\quad\quad k + 1 < 2^{\boxed{}}$

We have shown for all natural numbers k, $S_k \to S_{k+1}$. This completes the Induction Step. It and the Basis Step tell us that proof is complete.

Section 11.5 Combinatorics: Permutations

In order to study probability, it is first necessary that we learn about **combinatorics**, the theory of counting. The study of permutations involves *order* and *arrangements*.

Example 1   How many 3-letter code symbols can be formed with the letters A, B, C *without* repetition (that is, using each letter only once)?

A	☐	C	ABC
	C		ACB
	A	C	☐
☐	☐	A	
	A		BCA
C	☐	☐	CAB
		A	☐

6 Possibilities

The set of all permutations is $\{ABC, \boxed{}, BAC, BCA, CAB, CBA\}$.

The Fundamental Counting Principle

Given a combined action, or *event*, in which the first action can be performed in n_1 ways, the second action can be performed in n_2 ways, and so on, the total number of ways in which the combined action can be performed is the product

$$n_1 \cdot n_2 \cdot n_3 \cdot \ \cdots \ \cdot n_k.$$

Example 2   How many 3-letter code symbols can be formed with the letters A, B, C, D, E *with* repetition (that is, allowing letters to be repeated)?

$5 \cdot 5 \cdot \boxed{} = \boxed{}$

Thus, there are $\boxed{}$ code symbols.

Permutations

A **permutation** of a set of n objects is an ordered arrangement of all n objects.

The total number of permutations of n objects, denoted $_nP_n$ is given by

$$_nP_n = n(n-1)(n-2)\cdots 3\cdot 2\cdot 1.$$

Example 3   Find each of the following.

a) $_4P_4 = 4\cdot \boxed{}\cdot 2\cdot 1 = \boxed{}$

b) $_7P_7 = \boxed{}\cdot 6\cdot \boxed{}\cdot 4\cdot 3\cdot 2\cdot 1 = \boxed{}$

Example 4   In how many ways can 9 packages be placed in 9 mailboxes, one package in a box?

$$_nP_n = n(n-1)(n-2)\cdots 3\cdot 2\cdot 1$$

$$_9P_9 = 9\cdot \boxed{}\cdot 7\cdot 6\cdot \boxed{}\cdot 4\cdot \boxed{}\cdot 2\cdot 1$$

$$= \boxed{}$$

Factorial Notation

- For any natural number n, $n! = n(n-1)(n-2)\cdots 3\cdot 2\cdot 1.$

- For any natural number n, $n! = n(n-1)!.$

- For the number 0, $0! = 1.$

- $_nP_n = n!$

Example 5   Rewrite 7! with a factor of 5!.

$$7! = 7\cdot \boxed{} = 7\cdot 6\cdot \boxed{}$$

The Number of Permutations of n Objects Taken k at a Time

The number of permutations of a set of n objects taken k at a time, denoted $_nP_k$, is given by

$$_nP_k = \underbrace{n(n-1)(n-2)\cdots\left[n-(k-1)\right]}_{k \text{ factors}} \qquad (1)$$

$$= \frac{n!}{(n-k)!}. \qquad (2)$$

Example 6   Compute $_8P_4$ using both forms of the formula.

$$_8P_4 = 8 \cdot \boxed{} \cdot 6 \cdot 5 = \boxed{}$$

$$_8P_4 = \frac{8!}{(8-4)!} = \frac{8!}{\boxed{}}$$

$$= \frac{8 \cdot 7 \cdot 6 \cdot 5 \cdot \boxed{}}{4!}$$

$$= 8 \cdot 7 \cdot 6 \cdot 5 = \boxed{}$$

Example 7   *Flags of Nations.* The flags of many nations consist of three horizontal stripes. For example, the flag of Ireland has its first stripe green, its second white, and its third orange. Suppose that the following 9 colors are available:

   {black, yellow, red, blue, white, gold, orange, pink, purple}.

How many different flags of three horizontal stripes can be made without repetition of colors in a flag? (This assumes that the order in which the stripes appear is considered.)

$$_9P_3 = 9 \cdot 8 \cdot \boxed{} = \boxed{}$$

Example 8   *Batting Orders.* A baseball manager arranges the batting order as follows: The 4 infielders will bat first. Then the 3 outfielders, the catcher, and the pitcher will follow, not necessarily in that order. How many different batting orders are possible?

   4 Infielders $\boxed{}$ other players

$$_4P_4 \cdot {}_5P_5 = 4! \boxed{} = \boxed{}$$

The number of distinct arrangements of n objects taken k at a time, allowing repetition, is n^k.

Example 9   How many 5-letter code symbols can be formed with the letters A, B, C, and D if we allow a letter to occur more than once?

$$4 \cdot \boxed{} \cdot 4 \cdot 4 \cdot 4 = 4^{\boxed{}}$$

$$= \boxed{}$$

For a set of n objects in which n_1 are of one kind, n_2 are of another kind, ..., and n_k are of a kth kind, the number of distinguishable permutations is

$$\frac{n!}{n_1! \cdot n_2! \cdot \ldots \cdot n_k!}.$$

Example 10   In how many distinguishable ways can the letters of the word CINCINNATI be arranged?

2 C's   3 I's   3 N's   1 A   1 T

$$N = \frac{\boxed{}}{2!\boxed{} \cdot 3! \cdot 1! \cdot 1!}$$

$$= 50,400$$

The letters of the word CINCINNATI can be arranged in 50,400 distinguishable ways.

Section 11.6 Combinatorics: Combinations

We sometimes make a selection from a set *without regard to order*. Such a selection is called a *combination*.

Example 1   Find all the combinations of 3 letters taken from the set of 5 letters $\{A, B, C, D, E\}$.

$\{A, B, C\}$ $\{A, B, \boxed{}\}$ $\{A, B, \boxed{}\}$

$\{A, \boxed{}, D\}$ $\{A, C, E\}$ $\{A, \boxed{}, E\}$

$\{\boxed{}, C, D\}$ $\{B, C, \boxed{}\}$ $\{B, D, E\}$

$\{C, \boxed{}, E\}$

There are 10 combinations of the 5 letters taken 3 at a time.

Combinations of *n* Objects Taken *k* at a Time

The total number of combinations of *n* objects taken *k* at a time, denoted $_nC_k$, is given by

$$_nC_k = \binom{n}{k} = \frac{n!}{k!(n-k)!}, \qquad (1)$$

or

$$_nC_k = \binom{n}{k} = \frac{_nP_k}{k!} = \frac{n(n-1)(n-2)\cdots[n-(k-1)]}{k!}. \qquad (2)$$

Example 2   Evaluate $\binom{7}{5}$, using forms (1) and (2).

$$\binom{7}{5} = \frac{7!}{5!(\boxed{})!} = \frac{7!}{5!\,2!} = \frac{7\cdot 6 \cdot \boxed{}}{5!\,2!}$$

$$= \frac{7\cdot 6}{\boxed{}\cdot 1} = \boxed{}$$

$$\binom{7}{5} = \frac{7\cdot 6\cdot 5\cdot 4\cdot \boxed{}}{5\cdot 4\cdot 3\cdot \boxed{}\cdot 1}$$

$$= \frac{7\cdot 6}{2\cdot 1} = \boxed{}$$

Example 3   Evaluate $\begin{pmatrix} n \\ 0 \end{pmatrix}$ and $\begin{pmatrix} n \\ 2 \end{pmatrix}$.

$$\begin{pmatrix} n \\ 0 \end{pmatrix} = \frac{n!}{\boxed{}(n-0)!} = \frac{n!}{1 \cdot \boxed{}} = \boxed{} \qquad \text{Using form 1}$$

$$\begin{pmatrix} n \\ 2 \end{pmatrix} = \frac{n(\boxed{})}{2!} = \frac{n(n-1)}{\boxed{}}, \text{ or } \frac{n^2 - n}{2} \qquad \text{Using form 2}$$

Subsets of Size k and of Size $n - k$

$$\begin{pmatrix} n \\ k \end{pmatrix} = \begin{pmatrix} n \\ n-k \end{pmatrix} \quad \text{and} \quad {}_nC_k = {}_nC_{n-k}$$

The number of subsets of size k of a set with n objects is the same as the number of subsets of size $n - k$. The number of combinations of n objects taken k at a time is the same as the number of combinations of n objects taken $n - k$ at a time.

Example 4   *Michigan Lottery.* Run by the state of Indiana, Hoosier Lotto is a twice-weekly lottery game with jackpots starting at $1 million. For a wager of $1, a player can choose 6 numbers from 1 through 48. If the numbers match those drawn by the state, the player wins.

a) How many 6-number combinations are there?

$$_{48}C_6 = \begin{pmatrix} 48 \\ 6 \end{pmatrix}$$

$$= \frac{48!}{\boxed{}(48-6)!} = \frac{48!}{6! \boxed{}}$$

$$= \frac{48 \cdot 47 \cdot 46 \cdot 45 \cdot 44 \cdot 43 \cdot \boxed{}}{6 \cdot 5 \cdot 4 \cdot 3 \cdot 2 \cdot 1 \cdot 42!}$$

$$= \frac{48 \cdot 47 \cdot 46 \cdot 45 \cdot 44 \cdot 43}{6 \cdot 5 \cdot \boxed{} \cdot 3 \cdot 2 \cdot 1}$$

$$= \boxed{}$$

b) Suppose it takes you 10 min to pick your numbers and buy a game ticket. How many tickets can you buy in 4 days?

$$4 \text{ days} = 4 \text{ days} \cdot \frac{\boxed{}}{1 \text{ day}} \cdot \frac{\boxed{}}{1 \text{ hr}}$$

$$= \boxed{} \text{ min}$$

$$\frac{5760}{10} = \boxed{} \text{ tickets}$$

c) How many people would you have to hire for 4 days to buy tickets with all the possible combinations and ensure that you win?

$$\frac{12,271,512}{\boxed{}} = \boxed{} \text{ people}$$

Example 5   How many committees can be formed from a group of 5 governors and 7 senators if each committee consists of 3 governors and 4 senators?

$$_nC_k = \frac{n!}{k!(n-k)!}$$

$$_5C_3 \cdot _7C_4 = \frac{5!}{\boxed{}(5-3)!} \cdot \frac{7!}{\boxed{}(7-4)!}$$

$$= \frac{5 \cdot 4 \cdot \boxed{}}{3! \cdot 2 \cdot 1} \cdot \frac{7 \cdot 6 \cdot 5 \cdot \boxed{}}{4! \cdot 3 \cdot 2 \cdot 1}$$

$$= \frac{5 \cdot 2 \cdot 2}{\boxed{}} \cdot \frac{7 \cdot 6 \cdot 5}{\boxed{}}$$

$$= 5 \cdot 2 \cdot 7 \cdot 5 = 10 \cdot \boxed{} = 350$$

There are $\boxed{}$ possible committees.

Section 11.7   The Binomial Theorem

Pascal's Triangle

$$
\begin{array}{ccccccccccccc}
 & & & & & & 1 & & & & & & \\
 & & & & & 1 & & 1 & & & & & \\
 & & & & 1 & & 2 & & 1 & & & & \\
 & & & 1 & & 3 & & 3 & & 1 & & & \\
 & & 1 & & 4 & & 6 & & 4 & & 1 & & \\
 & 1 & & 5 & & 10 & & 10 & & 5 & & 1 & \\
1 & & 6 & & 15 & & 20 & & 15 & & 6 & & 1
\end{array}
$$

This pattern continues. Note that there are always 1's on the outside, and each remaining number is the sum of the two numbers above it.

The Binomial Theorem Using Pascal's Triangle

For any binomial $a+b$ and any natural number n,

$$(a+b)^n = c_0 a^n b^0 + c_1 a^{n-1} b^1 + c_2 a^{n-2} b^2 + \ldots + c_{n-1} a^1 b^{n-1} + c_n a^0 b^n,$$

where the numbers $c_0, c_1, c_2, \cdots, c_{n-1}, c_n$ are from the $(n+1)$st row of Pascal's triangle.

Example 1   Expand: $(u-v)^5$.

$$(a+b)^n \qquad a = u \qquad b = -v \qquad n = \boxed{}$$

$$(u-v)^5 = \left[u + \left(\boxed{} \right) \right]^5$$

We use the $\boxed{}$ row of Pascal's triangle.

$$1 \quad 5 \quad 10 \quad \boxed{} \quad 5 \quad 1$$

$$(u-v)^5 = 1(u)^5 + \boxed{}(u)^4(-v)^1 + 10(u)^{\boxed{}}(-v)^2$$

$$+ \boxed{}(u)^2(-v)^3 + 5(u)\left(\boxed{}\right)^4 + 1(-v)^{\boxed{}}$$

$$= u^5 \boxed{} 5u^4 v \boxed{} 10u^3 v^2 - 10u^2 v^3 + 5uv^4 - v^5$$

Example 2   Expand: $\left(2t + \dfrac{3}{t}\right)^4$.

$$(a+b)^n \qquad a = 2t \qquad b = \dfrac{3}{t} \qquad n = 4$$

We use the $\boxed{}$ row of Pascal's triangle.

$$1 \quad \boxed{} \quad 6 \quad 4 \quad 1$$

$$\left(2t + \dfrac{3}{t}\right)^4 = 1(2t)^4 + 4(2t)^{\boxed{}}\left(\dfrac{3}{t}\right)^1 + \boxed{}(2t)^2\left(\dfrac{3}{t}\right)^2$$

$$+ 4(2t)^1\left(\boxed{}\right)^3 + 1\left(\dfrac{3}{t}\right)^4$$

$$= 1\left(\boxed{}\right) + 4(8t^3)\left(\dfrac{3}{t}\right) + 6(4t^2)\left(\boxed{}\right) + 4\left(\boxed{}\right)\left(\dfrac{27}{t^3}\right) + 1\left(\dfrac{81}{t^4}\right)$$

$$= 16t^4 + \boxed{} \cdot t^2 + \boxed{} + 216t^{\boxed{}} + 81t^{-4}$$

The Binomial Theorem Using Combination Notation

For any binomial $a+b$ and any natural number n,

$$(a+b)^n = \binom{n}{0}a^n b^0 + \binom{n}{1}a^{n-1}b^1 + \binom{n}{2}a^{n-2}b^2 + \cdots + \binom{n}{n-1}a^1 b^{n-1} + \binom{n}{n}a^0 b^n$$

$$= \sum_{k=0}^{n}\binom{n}{k}a^{n-k}b^k.$$

Example 3   Expand: $\left(x^2 - 2y\right)^5$.

$$\left(a+b\right)^n \quad a = x^2 \quad b = \boxed{} \quad n = 5$$

$$\left(x^2 - 2y\right)^5 = \binom{5}{\boxed{}}\left(x^2\right)^5 + \binom{5}{1}\left(x^2\right)^{\boxed{}}\left(-2y\right)^1$$

$$+ \binom{5}{2}\left(\boxed{}\right)^3\left(-2y\right)^2 + \binom{5}{\boxed{}}\left(x^2\right)^2\left(-2y\right)^{\boxed{}}$$

$$+ \binom{5}{4}\left(x^2\right)^1\left(-2y\right)^4 + \binom{5}{\boxed{}}\left(-2y\right)^5$$

$$= \frac{5!}{0!\,5!}x^{\boxed{}} + \frac{5!}{1!\,\boxed{}}x^8\left(-2y\right) + \frac{5!}{\boxed{}\,3!}x^6\left(4y^2\right)$$

$$+ \frac{5!}{3!\,2!}x^4\left(-8y^3\right) + \frac{5!}{4!\,1!}\boxed{}\left(16y^4\right) + \frac{5!}{5!\,0!}\left(\boxed{}\right)$$

$$= 1x^{10} + \boxed{} \cdot x^8\left(-2y\right) + 10 \cdot x^6\left(4y^2\right)$$

$$+ \boxed{} \cdot x^4\left(-8y^3\right) + 5x^2\left(16y^4\right) + 1\left(-32y^5\right)$$

$$= x^{10} \boxed{} 10x^8 y \boxed{} 40x^6 y^2 - 80x^4 y^3 + 80x^2 y^4 \boxed{} 32y^5$$

Example 4   Expand: $\left(\dfrac{2}{x} + 3\sqrt{x}\right)^4$.

$$\left(a+b\right)^n \quad a = \frac{2}{x} \quad b = 3\sqrt{x} \quad n = 4$$

$$\left(\frac{2}{x} + 3\sqrt{x}\right)^4 = \binom{4}{0}\left(\frac{2}{x}\right)^4 + \binom{4}{1}\left(\frac{2}{x}\right)^{\boxed{}}\left(3\sqrt{x}\right)$$

$$+ \binom{4}{\boxed{}}\left(\frac{2}{x}\right)^2\left(3\sqrt{x}\right)^2 + \binom{4}{3}\left(\frac{2}{x}\right)\left(\boxed{}\right)^3 + \binom{4}{4}\left(3\sqrt{x}\right)^4$$

$$= \frac{4!}{0!\,4!}\left(\frac{\boxed{}}{x^4}\right) + \frac{4!}{1!\,\boxed{}}\left(\frac{8}{x^3}\right)\left(3x^{\boxed{}}\right)$$

$$+ \frac{4!}{2!\,2!}\left(\frac{4}{\boxed{}}\right)\left(9x\right) + \frac{4!}{3!\,1!}\left(\frac{2}{x}\right)\left(27x^{\boxed{}}\right) + \frac{4!}{4!\,0!}\left(\boxed{}x^2\right)$$

$$= \frac{16}{x^4} + \frac{96}{x^{\boxed{}}} + \frac{216}{\boxed{}} + 216x^{\boxed{}} + 81x^2$$

Finding the $(k+1)$st Term

The $(k+1)$st term of $(a+b)^n$ is $\binom{n}{k} a^{n-k} b^k$.

Example 5 Find the 5th term in the expansion of $(2x - 5y)^6$.

$$\binom{n}{k} a^{n-k} b^k \qquad a = 2x \qquad b = -5y$$

$$n = \boxed{}$$

$$5 = 4 + 1$$

$$k = \boxed{}$$

5th term is $\quad \binom{6}{4}\left(\boxed{}\right)^{6-4}(-5y)^{\boxed{}}$

$$= \frac{6!}{4!\,\boxed{}}(2x)^2(-5y)^4 = \boxed{}\, x^2 y^4$$

Example 6 Find the 8th term in the expansion of $(3x - 2)^{10}$.

$$\binom{n}{k} a^{n-k} b^k \qquad a = 3x \qquad b = \boxed{}$$

$$n = 10$$

$$8 = 7 + 1$$

$$k = \boxed{}$$

8th term is $\quad \binom{10}{7}(3x)^{10-\boxed{}}\left(\boxed{}\right)^7$

$$= \frac{10!}{7!\,\boxed{}}(3x)^3(-2)^7 = \boxed{}\, x^3$$

Total Number of Subsets

The total number of subsets of a set with n elements is 2^n.

Example 7 The set $\{A, B, C, D, E\}$ has how many subsets?

Number of subsets is $2^{\boxed{}}$, or $\boxed{}$.

Example 8   Wendy's, a national restaurant chain, offers the following toppings for its hamburgers:

{catsup, mustard, mayonnaise, tomato, lettuce, onions, pickle}.

In how many different ways can Wendy's serve hamburgers, excluding size of hamburger or number of patties?

$$\binom{7}{0} + \binom{7}{1} + \binom{7}{2} + \ldots + \binom{7}{7} = 2^{\boxed{}} = \boxed{}$$

Wendy's serves hamburgers in $\boxed{}$ different ways.

Section 11.8 Probability

> **Principle P (Experimental)**
>
> Given an experiment in which n observations are made, if a situation, or event, E occurs m times out of n observations, then we say that the **experimental probability** of the event, $P(E)$, is given by
>
> $$P(E) = \frac{m}{n}.$$

Example 1   *Television Ratings.* There are an estimated 114,200,000 households in the United States that have at least one television. Each week, viewing information is collected and reported. One week, 28,510,000 households tuned in to the 2013 Grammy Awards ceremony on CBS, and 14,204,000 households tuned in to the action series "NCIS" on CBS (*Source:* Nielsen Media Research). What is the probability that a television household tuned in to the Grammy Awards ceremony during the given week? to "NCIS"?

$$P = \frac{28,510,000}{\boxed{}} \approx 0.2496 \approx \boxed{}\,\%$$

$$P = \frac{14,204,000}{\boxed{}} \approx 0.1244 \approx \boxed{}\,\%$$

Example 2   *Sociological Survey.* The authors of this text conducted an experimental survey to determine the number of people who are left-handed, right-handed, or both. The results are 82 right-handed, 17 left-handed, and 1 ambidextrous.

a) Determine the probability that a person is right-handed.

Right-handed	82
Left-handed	$\boxed{}$
Ambidextrous	1
Total	$\overline{100}$

$$P(\text{Right-handed}) = \frac{\text{Number right-handed}}{\text{Total}} = \frac{\boxed{}}{100}, \text{ or } 0.82, \text{ or } 82\%$$

b) Determine the probability that a person is left-handed.

$$P(\text{Left-handed}) = \frac{\text{Number left-handed}}{\text{Total}} = \frac{17}{100}, \text{ or } 0.17, \text{ or } \boxed{}\,\%$$

c) Determine the probability that a person is ambidextrous (uses both hands with equal ability).

$$P(\text{Both}) = \frac{\text{Number ambidextrous}}{\text{Total}} = \frac{1}{100}, \text{ or } \boxed{}, \text{ or } 1\%$$

d) There are 130 students signed up for tennis lessons in a summer program offered by a community school corporation. On the basis of the data in this experiment, how many of the students would you expect to be left-handed?

We would expect ▢ % to be left-handed.

$0.17 \cdot \boxed{} = 22.1$

We would expect about ▢ students to be left-handed.

Example 3   *Dart Throwing.* Consider a dartboard divided into 3 equal sections: red, black, and white. Assume that the experiment is "throwing a dart" and that the dart hits the board. Find each of the following.

a) The outcomes

　　Hitting red　　(R)

　　Hitting black　$\left(\boxed{}\right)$

　　Hitting white　(W)

b) The sample space

　　$\left\{ R, \boxed{}, B \right\}$

Example 4   *Die Rolling.* A die (pl., dice) is a cube, with six faces, each containing a number of dots from 1 to 6 on each side.

Suppose that a die is rolled. Find each of the following

a) The outcomes

　　$1, 2, 3, \boxed{}, 5, 6$

b) The sample space

　　$\left\{ 1, \boxed{}, 3, 4, 5, 6 \right\}$

Principle P (Theoretical)

If an event E can occur m ways out of n possible equally likely outcomes of a sample space S, then the **theoretical probability** of the event, $P(E)$, is given by

$$P(E) = \frac{m}{n}.$$

Example 5   Suppose that we select, without looking, one marble from a bag containing 3 red marbles and 4 green marbles. What is the probability of selecting a red marble?

$$P(\text{selecting a red marble}) = \frac{\boxed{}}{7}$$

Example 6   What is the probability of rolling an even number on a die?

1   2   3   4   5   6

$$P(\text{Rolling an even number}) = \frac{3}{\boxed{}} = \frac{1}{2}$$

Example 7   What is the probability of drawing an ace from a well-shuffled deck of cards?

A deck contains $\boxed{}$ cards.

There are $\boxed{}$ aces.

$$P(\text{drawing an ace}) = \frac{4}{52}, \text{ or } \frac{1}{\boxed{}}$$

Probability Properties

a) If an event E cannot occur, then $P(E) = 0$.

b) If an event E is certain to occur, then $P(E) = 1$.

c) The probability that an event E will occur is a number from 0 to 1: $0 \le P(E) \le 1$.

Example 8   Suppose that 2 cards are drawn from a well-shuffled deck of 52 cards. What is the probability that both of them are spades?

$$P(\text{drawing 2 spades}) = \frac{_{13}C_{\boxed{}}}{_{52}C_2}$$

$$= \frac{\boxed{}}{1326} = \frac{1}{\boxed{}}$$

Example 9   Suppose that 3 people are selected at random from a group that consists of 6 men and 4 women. What is the probability that 1 man and 2 women are selected?

$$P = \frac{_6C_1 \cdot {}_4C_2}{_{10}C_3} = \frac{3}{10}$$

Example 10   *Rolling Two Dice.* What is the probability of getting a total of 8 on a roll of a pair of dice?

There are $6 \cdot 6$, or 36 possible outcomes. Five of these give a total of 8.

$$P = \frac{\boxed{}}{36}$$